Fundamentals of Industrial Robots and Robotics

THE PWS-KENT SERIES IN TECHNOLOGY

FUNDAMENTALS OF INDUSTRIAL ROBOTS AND ROBOTICS

Rex Miller
State University College at Buffalo

 ■ PWS–KENT Publishing Company ■ Boston

PWS-KENT
Publishing Company

20 Park Plaza
Boston, Massachusetts 02116

PWS-KENT Publishing Company is a division of Wadsworth, Inc.

Library of Congress Cataloging-in-Publication Data
Miller, Rex, 1929–
    Fundamentals of industrial robots and robotics / Rex Miller.
      p.   cm.
    Bibliography: p.
    Includes index.
    ISBN 0-534-91470-5
    1. Robotics.   I. Title.
TJ211.M535 1988
629.8'92 — dc19   87-28438
                                     CIP

Printed in the United States of America
88 89 90 91 92 — 10 9 8 7 6 5 4 3 2 1

Editorial Assistant: Mary Thomas
Production: Technical Texts, Inc./Jean T. Peck
Interior Design: Sylvia Dovner
Cover Design: Sylvia Dovner
Cover Art: Slide Graphics of New England, Inc.
Typesetting: A&B Typesetters
Cover Printing: New England Book Components, Inc.
Printing and Binding: Arcata Graphics/Halliday

This book is dedicated to
John Byczkowski and Joseph Guetta.

PREFACE

The book was written with a number of purposes in mind. While many people want to know more about robots and robotics, most do not have engineering or technical backgrounds in pneumatics, hydraulics, and electronics to understand what the robot is all about internally or conceptually. Further, some people who do have the necessary background do not know where to start in looking at the future of the robot in their own trade or profession. Thus, *Fundamentals of Industrial Robots and Robotics* is intended as a comprehensive introduction to the topic.

The book is suitable for use in a first course on robotics for students in industrial electronics programs, as well as those in mechanical, manufacturing, or industrial technology. It is also designed to serve as a source of information on robots and robotics for robot hobbyists, professional machinists, electricians, and electronics technicians. It provides a broad view of the subject without overwhelming the reader with technical details or jargon.

The text relies upon the real world of robots to bring excitement to its pages. Up-to-date examples of industrial robots and practical applications are emphasized throughout the book. Ample illustrations are provided to clarify the discussion and to aid the reader in recognizing robot parts and movements. The end-of-chapter definitions and a comprehensive glossary are included to make terms applicable to robots easy to understand and master.

The organization of the book is flexible and allows for individual preferences in the order of study. Chapter 1 provides an overview of robotics. It includes a definition of what constitutes a robot and outlines both the positive and negative aspects of robots, as well as their interaction with human labor. Chapter 1 also contains a brief review of computer programs, languages, and microprocessors.

Chapter 2 identifies various types of robots. Parts of the robot and robotic motion capabilities are also examined. Chapter 3 covers the mechanical components of robots such as drive systems, pumps, and motors. Sensor types and sensing capabilities are discussed in Chapter 4.

Chapter 5 covers control methods for robots, including various methods of robot programming, and Chapter 6 examines the topic of robot and computer interfacing. Chapter 7 discusses different uses for robots in industry and the future of robots and robotics. Chapter 8 provides a list of selected manufacturers and equipment, along with selected specifications, descriptive information, and illustrations from the manufacturers' catalogs.

In addition to a glossary, two appendixes are included to add to the book's usefulness. Appendix I provides a cross-comparison table of specifications of 166 different robot models from 33 manufacturers. Electronics and fluid power schematic symbols are illustrated in Appendix II.

Whether beginners or individuals who have worked with machines for some time, readers will gain not only fundamental knowledge but also new insights into the complex field of robotics. The section on putting the robot to work will give readers a clear idea of what these machines can and cannot do. While robots have a long way to go before they can do all the things we dream they will do, they are an exciting and dynamic force that must be seriously considered by everyone, no matter what their occupation or interests. The goal of this book is to provide the necessary information in such a way that it can be used effectively by readers of all skill levels and backgrounds.

Acknowledgments

No book is ever completed without the energy and efforts of many people. This one is no exception, and I would like to thank the many people, both named and unnamed here, whose contributions made this book a reality.

Throughout the various stages of writing the manuscript, helpful comments and suggestions were provided by a number of people: Professor Jimmy Hutto, MacArthur State Technical College, Opp, Alabama; Professor C. Eddie Gore, Carroll County Area Vocational–Technical School, Waco, Georgia; Professor A. F. Adkins, Amarillo College, Amarillo, Texas; Professor Michael W. Koehler, Jefferson Community College–Southwest, Louisville, Kentucky; Professor Patrick J. Niedbala, Schoolcraft College, Livonia, Michigan; Professor Ralph R. Stockard, Technical College of Alamance, Haw River, North Carolina; Professor Ronald G. Barker, Georgia State Department of Education, Atlanta, Georgia; Professor Mark E. Meyer, College of DuPage, Glen Ellyn, Illinois; and Professor Robert L. Hinders, Western Iowa Technical Community College, Sioux City, Iowa.

Many businesses were also most helpful in supplying information and illustrations so necessary to making the book worthwhile: Automatix, Inc.; Binks Manufacturing Company; Cincinnati Milacron/Industrial Robots Division; CAMCO, Commercial Cam Division, Emerson Electric Company; Compact Air Products, Inc.; Cybotech® Industrial Robots; Elicon®; ESAB North America, Inc.; Fared Robot Systems, Inc.; Feedmatic-Detroit, Inc.; Feedback, Inc.; GCA Corporation/Industrial Systems Group; Graymark International, Inc.; Hobart Brothers Company; International Business Machines Corporation/Manufacturing Systems Products Division; International Robomation/Intelligence; I.S.I. Manufacturing, Inc.; Mack Corporation; Microbot®; Microswitch, a Honeywell Division; New Jersey Zinc Company; Pick-O-Matic Systems®; Prab Robots, Inc.; RCA, Radio Corporation of America; Rhino® Robots, Inc.; Schrader-Bellows, Division of Parker-Hannifin; Seiko Instruments USA, Inc.; Thermwood Robotics Division, Thermwood Corporation; Unimation Incorporated, a Westinghouse Company; Warner Clutch and Brake Company; and Yaskawa Electric America, Inc.

CONTENTS

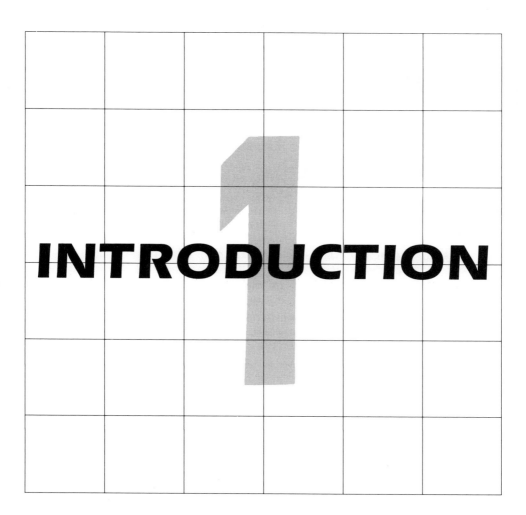

INTRODUCTION

R obots have captured the imagination of writers and movie producers for some time. Only recently have they become useful in the production of quality products, and it is here that the greatest amount of time and effort is being spent in robot development today. The ability to produce quality products is of utmost importance since the consumer benefits and the manufacturer stays in business.

☐ ROBOT DEFINITION

What is a robot? There are a number of definitions, but a simple definition that serves our purpose here is: A robot is a reprogrammable, multifunctional manipulator designed to move material, parts, tools, or specialized devices through variable programmed motions for the performance of a variety of tasks.

A robot can be classified as a system that simulates human activities from computer instruction. See Figure 1–1. These definitions allow us to take a closer look at the system that will produce the actions needed for a robot to perform tasks that humans can do. The computer is the secret to the system since it is, in effect, the brain of the device or system. The computer is an integral part of any robot and must be taken into consideration whenever the robot is studied as a device, a system, or a means of eliminating human effort.

Keep in mind that some nonintelligent robots do not use electronics for brains. A lot of pick-and-place robots are cam controlled. See Figure 1–2. They simply pick up their load and place it elsewhere. Loading and unloading tasks usually are performed by this type of robot.

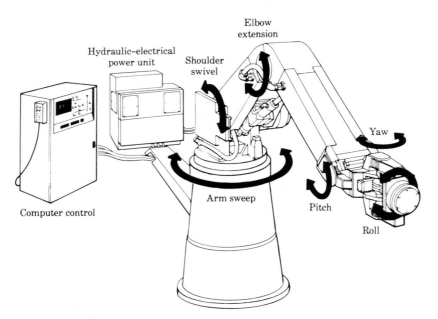

FIGURE 1–1   Complete industrial robot system (courtesy of CINCINNATI MILACRON)

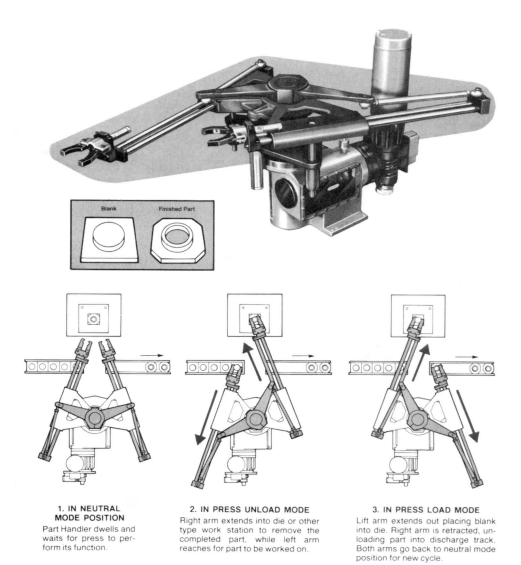

1. IN NEUTRAL MODE POSITION
Part Handler dwells and waits for press to perform its function.

2. IN PRESS UNLOAD MODE
Right arm extends into die or other type work station to remove the completed part, while left arm reaches for part to be worked on.

3. IN PRESS LOAD MODE
Lift arm extends out placing blank into die. Right arm is retracted, unloading part into discharge track. Both arms go back to neutral mode position for new cycle.

FIGURE 1-2 Pick-and-place robot used to load a press (courtesy of PICKOMATIC SYSTEMS®)

☐ ROBOT HISTORY

A bit of history will place the robot into perspective and help explain its popularity. The robot is a relatively recent development.

Science fiction has featured robots for as long as they have been around, but it was in 1921 when Karl Capek wrote *R.U.R.* that the term *robot* came into common usage. Capek's book and play introduced the world to the word *robot*. Since Capek was Czechoslovakian, he used the word *robota* to describe the machine that performed like humans but did not have the senses of humans. The term *robota* means slave labor and was reduced to *robot* in English.

What is a true robot? A few ideas must be taken into consideration when you answer this question. Some of them are:

— A robot is a device or system that is programmed by a human to perform humanlike acts.
— A robot may sense various conditions and react in a preprogrammed manner.
— A robot may be able to react to various conditions in terms of the five human senses: sight, hearing, smell, taste, and touch. *Sensors* are available that allow all of these senses to be inserted into a system.
— A robot is a system that can operate on its own without human supervision.
— A robot may make decisions by comparing information received from sensors and reacting in a preprogrammed way.

The invention of large-scale computers in the mid-1950s helped the robot to become a reality. Robots then became more popular with the advent of the personal computer or *microcomputer* in the early 1970s. By incorporating the computer into the robotic system, it was possible to create a unit that could move, talk, lift things, see where it was going, and know what it was feeling.

□ COMPUTER PROGRAMS

Special computer programs designed for specific jobs are used to control robots. Industrial robots, for instance, are designed to do a particular operation. This one operation may be done over and over again, but the robot, unlike humans, does not become fatigued or bored. A *program* is written to take into consideration the exact tasks to be performed. In some instances, it may take years to analyze the moves needed to perform a particular job. This information then has to be fed into a computer in terms that it understands, and the computer then sends signals to the robot so it will perform exactly as desired. See Figure 1–3, where two robots are used to load and unload. By coordinating the motions of the two robots, the idle time of machine A is kept to under one second. The alternate path capability built into the command module allows a variety of sensors to detect reject parts and place them in an alternate location.

A. Robot and teach pendant

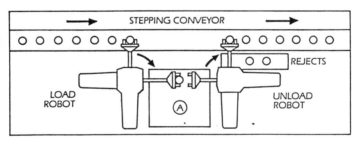

B. Loading and unloading

FIGURE 1-3 Robot with microprocessor and teach pendant especially designed to load and unload (courtesy of SCHRADER-BELLOWS)

Languages

Robots use a number of computer *languages*, which are designed for specific operations. Some of the languages used with robots are: AL (Stanford's Artificial Intelligence Lab language), VAL, AML (developed by IBM), Pascal, and ADA. These terms and more on languages and programming are discussed later in this book.

☐ MICROPROCESSORS

A *microprocessor* is just that. *Micro* means small and a *processor* is a device that can process things. In this case, it is used to process information fed to it from an external source.

We use the term today to describe a special purpose chip or portion of a chip that gets its instructions from a keyboard, joystick, mouse, or any number of sensors. See Figure 1–4. The chip can do math, make logical decisions, and work with words and symbols.

The robot usually has a dedicated microprocessor designed for its special job requirements. It may be a simple chip, or it may be a large mainframe computer.

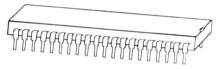

A. Microprocessor chip

B. Keyboard

C. Joystick

FIGURE 1–4 Microprocessor chip, keyboard for programming, and joystick

Without the microprocessor, the robot is limited in its application. With the microprocessor, it is possible to have the robot operate alone without connecting wires, other than for power purposes. More on the operation of microprocessors will be found later in this book.

☐ POSITIVE ASPECTS OF ROBOTS

In many instances, robots can perform work more efficiently than humans. They can work seven days a week, twenty-four hours a day, and thirty days a month without becoming bored or fatigued. The quality of their work can be checked and corrected immediately if found defective. Operating costs are low and downtime is minimal. Thousands of people will be needed in the future to design, repair, and install robots. New jobs will be created, and new training programs will have to be installed to improve the utilization of robots. Figure 1–5 shows how a robot is

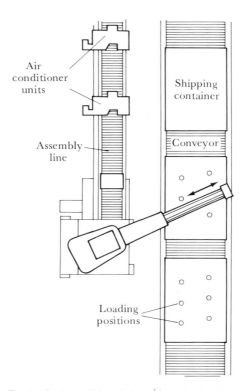

FIGURE 1–5 Robot used to pick up finished air conditioning units and pack them in shipping containers (from Malcolm *Robotics*)

used to pick up air conditioning units from an assembly line and place them into shipping cartons — a job that would quickly tire and bore human workers.

☐ NEGATIVE ASPECTS OF ROBOTS

Robots replace humans in the labor force. They require a higher level of maintenance than do most existing jobs. Therefore, they require retraining or replacement of humans now employed. The initial cost of robots is excessive for small firms. The technology is relatively untested at this time, and downtime is expensive.

☐ ROBOTS, HARD AUTOMATION, AND HUMAN LABOR

For many years the "American Way" was the best way. This applied to manufactured goods, standard of living, and everything else that we considered a part of the American Way. The United States was a large manufacturing nation that reached its level of operation during World War II. The pent-up demand for consumer goods after World War II presented a challenge to the manufacturing system as it then existed. More machines were made and more people were employed to meet the demand for industrial products. As demand slackened, emphasis on product quality increased. People demanded quality instead of quantity. This became evident during the oil crisis in 1974. Smaller cars were imported from Japan to fill the need for more fuel-efficient vehicles. As more and more Americans began driving Japanese cars, they noticed the quality of the product and demanded the same of their American counterparts.

As the quality of American life improved, more money was needed to support it. This meant that workers demanded more money to meet their expectations. Organized labor demanded, on behalf of the worker, more and more until the point was reached when it was no longer feasible to manufacture certain products in this country. Foreign companies were able to meet the demand for consumer products at lower prices since their labor and manufacturing costs were lower. American manufacturers began to look for ways to improve quality and reduce cost per item produced so they could effectively compete with foreign manufacturers.

Inasmuch as robots have many advantages over human labor, it was only natural that manufacturers looked in that direction to satisfy their labor needs. Figure 1-6 shows how the robot can do routine repetitive operations without tiring. They do, however, occasionally break down.

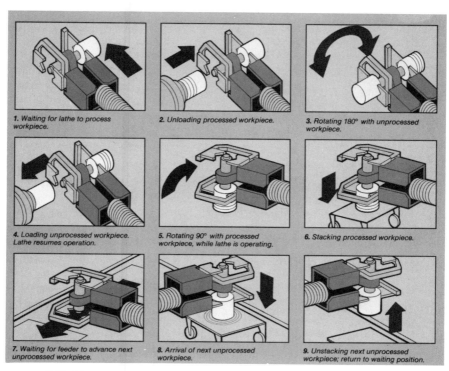

1. Waiting for lathe to process workpiece.

2. Unloading processed workpiece.

3. Rotating 180° with unprocessed workpiece.

4. Loading unprocessed workpiece. Lathe resumes operation.

5. Rotating 90° with processed workpiece, while lathe is operating.

6. Stacking processed workpiece.

7. Waiting for feeder to advance next unprocessed workpiece.

8. Arrival of next unprocessed workpiece.

9. Unstacking next unprocessed workpiece; return to waiting position.

FIGURE 1-6   Robot gripper at work (courtesy of CAMCO, COM-MERCIAL CAM DIVISION, EMERSON ELECTRIC CO.)

Robots will work in unpleasant locations. Health hazards are not of concern to the robot. Special safety equipment is not required for the robot to spray paint, weld, or handle dangerous chemicals. All this adds up to reduced production costs. As the day progresses, the tired worker has a tendency to pay less attention to details. This inattention results in a finished product of lower quality. This is especially noticeable in automobiles where spray paint can run or sag and weld joints are not made perfectly. The panels on the car may not be aligned, and the finished product may not operate properly, resulting in a very unsatisfied customer.

Robots, on the other hand, do not tire or change their work habits unless programmed to do so. They maintain the same level of operation throughout the day. With the utilization of robots, it is possible for American manufacturers to compete against lower labor costs in foreign countries. The initial investment is the only problem. After the initial investment, the overall operation of the production line is reduced or held constant. Small robots (Figure 1-7) are used to retrain humans to operate and maintain robots.

There are some advantages to human labor. Humans can start to

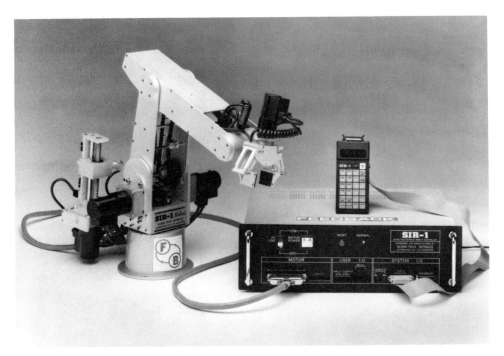

FIGURE 1–7   Educational robot – a complete unit for retraining people to work on robots (courtesy of FEEDBACK, INC.)

work immediately upon learning the task. Humans can also be laid off if they are no longer needed. There are, of course, some costs such as unemployment compensation and severance pay in terminations. If a company has very little money to invest in new equipment, using human labor is usually better. Robots are expensive initially.

Automated machinery will perform various operations with some degree of quality and dependability. However, if a design changes, it becomes expensive to replace the old equipment and buy new. Automated machinery is the best choice if you are going to produce a product for a long time. The investment in *hard automation* is warranted in such a case. However, if the design of the product is subject to quick change, it is best to choose robots to do the work. Robots are flexible and can be reprogrammed to do something else.

Return on investment is the primary consideration when deciding whether to use human labor, hard automation, or robots. The robot is only part of an automated system and probably the cheapest part at that. You have to have an in-feed device, an out-feed device (or parts delivery and removal system) as well as end-of-arm tooling, robot envelope security, and additional add-ons. All this has to be part of the equation when you ask: Will the robot pay for itself in five years? Average costs can be compared here for the sake of discussion only. The cost of an industrial robot is about $70,000, whereas the cost of an automated machine designed and installed is about $225,000. Human labor, including its benefit package, will average about $30,000 a year. As far as production time is concerned, the automated machine will produce the product fastest. The robot is next in terms of time needed to produce a quality part, with the human coming in last in terms of time. The return on investment must be considered by the person making the decision to go to robots, automated machines, or human labor.

If the money invested is tied up for more than five years, it is considered too long. The chief advantage of the robot in this consideration and choice is its flexibility. If the product being made changes, the robot can be reprogrammed to accommodate the design change. It usually takes humans much longer to be retrained. Figure 1–7 is an example of a retraining system. Time is money when it comes to manufacturing or production of anything. The quicker the line can be retooled, the more money that can be made and the better the return on investment.

☐ ROBOTS AND HUMANS

There are advantages and disadvantages to using robots. Humans also have favorable characteristics when it comes to using them in the manufacture of various products. One of the main advantages of human labor is that a person can be laid off if production needs change or the economy falters. The robot cannot be laid off. It goes on costing every day it is there, whether it is being used or not. It is an investment and must be considered as any other investment. The robot has to be paid for once the contract for its purchase has been issued.

Robots versus Humans

The robot-versus-human question becomes a major consideration if robots are installed in great numbers and replace many people in a particular location. This can cause turmoil in the job market. The number of robots in use today is very small in respect to what is expected in ten to twenty years. If 20,000 people are employed by a particular corporation, the installation of a few robots will not cause too much concern. If a new plant is built and robots are installed during construction, there is still very little in the way of labor problems.

However, estimates of the number of U.S. workers that will be displaced by robots range from 250,000 to 2 million within ten years. This can accelerate or diminish according to the economy. Such displacement of workers may create problems with the American work force. There is much speculation as to how the rapid deployment of robots will affect the social order of American industry. Many workers will have to be retrained for higher level jobs, creating a demand for more people to be employed in retraining programs. The transition period will likely last ten to fifteen years. During this time, it will be necessary to absorb displaced workers in training programs or other types of industries. American service industries are growing rapidly, and many displaced workers will be absorbed in this type of work.

The main disadvantage of robots is downtime. When a robot breaks down, an entire plant's production schedule may be affected, causing problems with sales and distribution as well as production. One machine off-line in a production line can result in many more being made nonproductive down the line. You must maintain a proper inventory of repair parts and employ a properly trained technician to keep the equipment operating at peak efficiency. A replacement robot may be kept ready (an expensive option), or a stock of major components must be kept on hand.

Competent personnel must also be available to work quickly and effectively. In summary, robots, like other equipment, offer advantages and disadvantages.

Each advantage and disadvantage must be considered in respect to the robot's intended use before a decision to acquire is made. Most industries that have investigated the use of robots have found that the advantages outweigh the disadvantages, and as a result more robots will be making more things in the future.

☐ INDUSTRIAL ROBOT APPLICATIONS

No doubt about it, robots are here to stay. A good example of a robot being put to use is shown in Figure 1–8, a basic robot system used for spray painting. It consists of a manipulator, control console, and hydraulic power supply. Electric-powered robots for painting can be dangerous since the fumes can easily be ignited by a spark.

Other examples of how industrial robots are used include the following applications.

- Machine loading and unloading: Placing parts where they are needed for machining or shipping.
- Materials handling: Packing parts or moving pallets.
- Fabrication: For making investment castings, grinding operations, and deburring; for water jet cutting, wire harness manufacture, applying glues, sealers, putty, and caulks; for drilling, fettling, and routing.
- Spray painting: Painting cars, furniture, and other objects.
- Welding: Welding cars, furniture, and steel structures.
- Assembly: Electronics, automobiles, and small appliances.
- Inspection and testing: Quality control looking for surface and interior defects, using vision sensors and feelers.

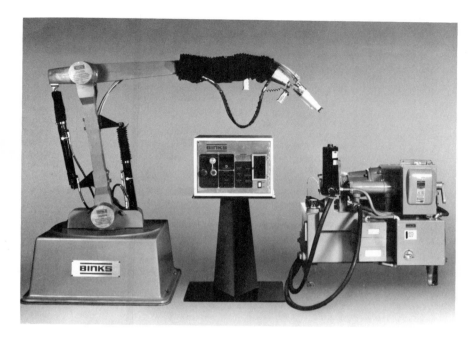

FIGURE 1-8   Spray painting robot – a basic robot system consisting of a six-axes manipulator, control console, and hydraulic power supply (courtesy of BINKS MANUFACTURING COMPANY)

□ SUMMARY

A robot is defined as a programmable, multifunctional, manipulator designed to move material, parts, tools, or specialized devices though variable programmed motions for the performance of a variety of tasks. The computer is the brain of the robot.

The robot is a relatively recent development. In 1921, Karl Capek used the Czech word *robota* in a book and play called *R.U.R.* The word was changed to *robot* in English.

A number of things must be considered when determining the answer to the question "what is a true robot?"

It is a device or system that is programmed by a human to perform humanlike acts.

It is a device or system that may sense various conditions and react in a preprogrammed manner.

It may be able to react to various conditions in terms of the five human senses.

It is a system that can operate on its own without any human supervision.

It may make decisions by comparing information received from sensors and reacting in a preprogrammed way. These sensors are classified as magnetic, light-activated, heat-activated, and pressure-activated.

The robot became a reality in the mid-1950s. The development of the robot is closely tied to the development of the computer.

Special programs are used to control robots. They are designed for specific jobs and are written in special programming languages. Many languages exist for the control of robots.

The microprocessor is the brain of the robot. It has the ability to take sense signals and make the robot react in a planned way. The microprocessor is an electronic device made from silicon chips.

Robots can work seven days a week without a break. They are capable of performing tasks more efficiently than humans.

Robots are expensive and need highly trained technicians to keep them operational. They replace humans but create a demand for more highly skilled workers to keep them operating.

Robots are tied to the improvement of quality of manufactured products. There are advantages and disadvantages to the use of robots. Each advantage must be weighed against the disadvantage before making a decision to buy robots instead of using human labor.

Robots have the advantage of being retrained rather quickly. They are flexible and can be used to do more than one thing with a minimum of reprogramming or retraining.

The main disadvantage of a robot is downtime. If it breaks down, it may hold up an entire plant's production schedule.

Robots have a number of industrial applications that make them useful to larger manufacturers who can withstand the initial cost of the unit and its installation and debugging.

☐ KEY TERMS

hard automation   the use of the conventional assembly line method of producing a manufactured product with dedicated equipment

languages   a method of speaking to a robot (VAL, AL, AML, Pascal, and ADA are used. These languages are limited, precise, and rigid.)

microprocessor   a chip or part of a chip that has the ability to do math and react to various inputs from sensors; part of a robot – the brain

program   a sequence of commands instructing a robot to perform some task

programmable robot one that can be programmed or taught with a "teach box," a keyboard, or some input device

programmer a person who teaches the robot; a person who can communicate with the robot in its language

robot a system that simulates human activities from computerized instruction

sensor a device used to detect changes in temperature, light, pressure, sound, and other functions needed to make a robot aware of various conditions

☐ QUESTIONS

1. What is a robot?

2. When did robots become a reality?

3. Where does the word *robot* come from?

4. List five ways you can identify a true robot.

5. List at least five languages used by robots.

6. What is a microprocessor?

7. What are two positive aspects of robots?

8. What are two negative aspects of robots?

9. What is the difference between hard automation, robots, and human labor?

10. What is meant by return on investment?

11. What is the main advantage of the robot over humans as you see it?

12. List at least five operations that industrial robots can perform.

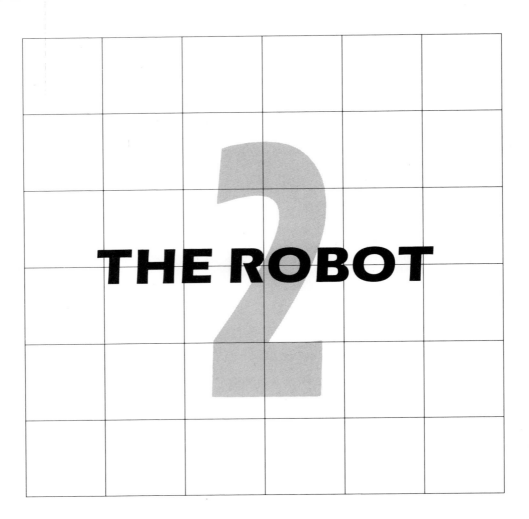

THE ROBOT

The robot is made up of a number of subsystems, which are usually standard and can be used for any number of other purposes. This makes the robot an inexpensive device to manufacture when compared to what it would cost if all the parts and systems were individually designed and handmade. The designer of a robot has a purpose in mind. This purpose may be to move an object from one place to another, or it may be a complicated maneuver requiring many subsystems to get the job done.

Robots can be classified in a number of ways. The classification system we use here is the end purpose of the robot; in other words, we classify the robots as industrial, laboratory, explorer, hobbyist, classroom, and entertainment. These are but a few of the jobs they do. We will concentrate on the broad category of industrial robots and introduce some of the others as the book develops.

☐ INDUSTRIAL ROBOTS

Industrial robots have arms with grippers attached. See Figure 2–1. The grippers are fingerlike and can grip or pick up various objects. They are used to pick and place. They pick up an object and place it elsewhere or move materials from one place to another. These robots can be programmed and computerized. The teach box is used to program the *microprocessor* used as the computer brain. Sensory robots, welding robots, and assembly robots usually have a self-contained microcomputer or minicomputer.

FIGURE 2–1   Industrial robot used to pick and place

☐ LABORATORY ROBOTS

Laboratory robots take many shapes and do many things. They are the beginnings of more sophisticated devices. They have microcomputer brains, multijointed arms, or advanced vision or tactile senses. Some have good hand-eye coordination and will eventually be used in industry to become more productive. Some of these may be mobile and others may be stationary. See Figure 2–2.

☐ EXPLORER ROBOTS

Explorer robots are used to go where humans cannot go or fear to tread. For instance, they are used in outerspace probes, to explore caves, dive far deeper underwater than humans can, and can be used to rescue people in sunken ships. The National Aeronautics and Space Administration (NASA) has done much to develop explorer robots to explore the surface of the moon and the surface of Mars. Explorer robots are sophisticated machines that have sensory systems and are remotely controlled or controlled by preprogrammed onboard computers. See Figure 2–3.

FIGURE 2–2   Laboratory robot used to handle dangerous materials (Ford Motor Company uses two of these to handle plastic instrument lenses.) (courtesy of SEIKO INSTRUMENTS USA, INC.)

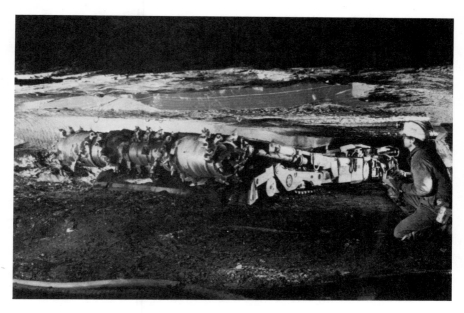

FIGURE 2–3   Explorer robot (courtesy of NATIONAL MINE SERV-ICE COMPANY)

☐ HOBBYIST ROBOTS

Most hobbyist robots are mobile. They are usually made to operate by rolling around on wheels propelled by small electric motors controlled by an onboard microprocessor. See Figure 2–4. The ultimate goal of most hobbyists is to have a housekeeper robot that will do the hard work of keeping the living quarters livable. Most hobbyist robots are equipped with speech-synthesis and speech-recognition systems. Some follow a line on the floor and others will follow preprogrammed instructions. Most have an arm or arms and resemble a person in appearance.

☐ CLASSROOM ROBOTS

Classroom robots are limited in application at this time. In the future the classroom robot will be able to move around the classroom and assist the teacher in various aspects of the teaching-learning process. Those now available in schools are mainly hobbyist in nature and are used to teach the fundamentals of robots to students engaged in learning practical applications for various electronic circuits.

Secondary schools, vocational institutions, junior colleges, and

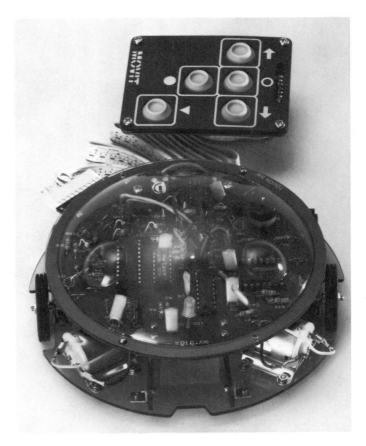

FIGURE 2-4   Hobbyist robot (courtesy of GRAYMARK INTERNA-
TIONAL, INC.)

universities are adding robotics courses to their curriculums in an effort
to prepare students to enter the work force for these advanced tech-
nologies.

Because industrial robots can cost from $50,000 to $100,000, are
intimidating to untrained workers, and are costly to repair, they are
highly impractical for training purposes. Rhino has developed the XR
series tabletop robotic system to meet the needs of manufacturing companies and educational institutions that require a high-quality, cost-
effective means of training. See Figure 2–5A. Microbot also makes an
educational robot that can be used to teach the basic functions of an
industrial robot. See Figure 2–5B.

For about one-tenth the cost of full-scale industrial robot systems, tabletop systems can totally simulate the larger systems, provid-

A. Complete robot training course using an IBM computer

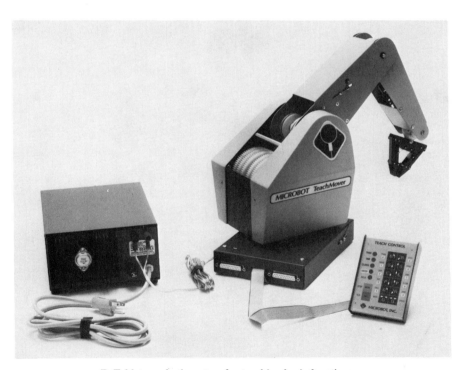

B. Tabletop robotic system for teaching basic functions

FIGURE 2–5 Classroom robots used to teach the fundamentals of robot operation (part A courtesy of RHINO® ROBOTS, INC.; part B courtesy of MICROBOT®)

ing cost-effective training that readily transfers to the industrial setting.

☐ ENTERTAINMENT ROBOTS

Entertainment robots are just beginning to be developed and made available. They have the ability to speak and respond to the spoken word. They can be used to entertain people at various events or operate as a roving advertisement. More uses for robots in the entertainment field will be forthcoming as more inexpensive sophisticated programming becomes available. For instance, look at Figure 2–6. The entertainment industry uses robots to take pictures where it would otherwise be impossible to photograph. Special cameras readily adaptable to robot control have been developed for better photography and television coverage of events and operations of all sorts.

FIGURE 2–6   Entertainment robot with a boom arm camera control system (courtesy of ELICON®)

□ THE MANIPULATOR

There are about 250 manufacturers of robots in the United States, Europe, and Japan, making it very difficult to identify all the parts used in the available robots. However, there are some common components that may be examined for a better perspective on how a robot works.

The *manipulator* is one of the three basic parts of the robot. The other two are the *controller* and the *power source*. See Figure 2–7. In order for the robot to do work each of these three components must be operational.

The manipulator is classified by certain arm movements. There are, for instance, four coordinate systems used to describe the arm movement: polar coordinates, cylindrical coordinates, Cartesian coordinates, and articulate (jointed-arm, spherical) coordinates.

Let's take a closer look at some of the robot components before we examine the coordinate systems. This will introduce you to the terms used in the later coordinate discussion.

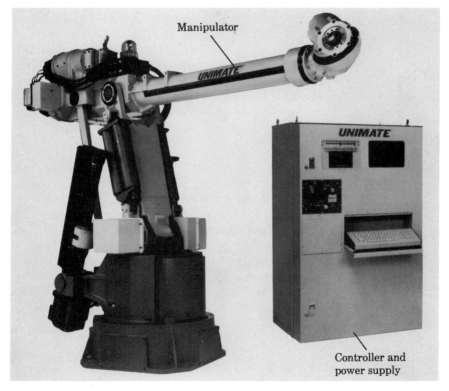

FIGURE 2–7 The three basic components of a robot (courtesy of UNIMATION INCORPORATED, a Westinghouse Company)

☐ BASE

The base of the robot is its anchor point. The base may be rigid. It is usually designed as a supporting unit for all the component parts of the robot. The base does not have to be stationary since it may become part of the operational requirements of the robot. It may be capable of any combination of motions, including rotation, extension, twisting, and linear. Most robots have the base anchored to the floor (Figure 2–8A), although, due to limited floorspace, they may be anchored to the ceiling or to suspended support systems overhead (Figure 2–8B). A track or conveyor system may be used to move the robot along as needed (Figure 2–8C).

☐ ARM

Some type of arm is found on most industrial robots. It may be jointed and resemble a human arm or it may be a slide-in/slide-out type used to grasp something and bring it back closer to the robot. A jointed arm consists of a base rotation axis, a shoulder rotation axis, and an elbow rotation axis. This type of arm provides the largest working envelope per area of floor space of any design thus far. If this is a six-axes type of arm, it requires some rather sophisticated computer control. See Figure 2–9. Most arms now have some type of joint. From one to six jointed arms may be attached to a single base for special jobs. The expense of controlling this type of movement is rather large because of its complexity.

☐ WRIST

Figure 2–10 shows how the wrist is attached to the jointed arm. The wrist is similar to a human wrist and can be designed with a wide range of motion, including extension, rotation, and twisting. This aids the robot in reaching places that are hard to reach by the human arm. It comes in handy especially when spray painting the interior of an automobile on the assembly line. It is also helpful in welding inside a pipe. This type of flexibility will improve the manufactured products we now enjoy and add others we were unable to fabricate earlier.

☐ GRIPPERS

The *grippers* are at the end of the wrist. They are used to hold whatever the robot is to manipulate. See Figure 2–11. Pick-and-place robots have grippers to move objects from one place to another.

A. Base mounted to the floor

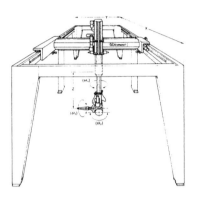

B. Base mounted on a gantry for moving over the work area from above

C. Base mounted on a track for movement of the whole unit

FIGURE 2-8 Robot bases (courtesy of GCA CORPORATION/INDUSTRIAL SYSTEMS GROUP)

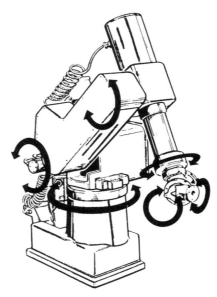

FIGURE 2-9   The six axes of a robot (courtesy of CINCINNATI MILACRON)

FIGURE 2-10   The three axes for the wrist of a manipulator arm (courtesy of CINCINNATI MILACRON)

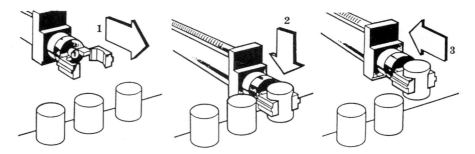

FIGURE 2-11   Robot arm (manipulator) with grippers for reaching out and picking up an object on a line and bringing it back to place it elsewhere (courtesy of CINCINNATI MILACRON)

Some robots have end-of-arm tooling instead of grippers. In such cases the robot is used primarily for one type of operation such as spray painting or welding. If a tool is attached, it is unnecessary to have a gripper on the end of the arm. A pneumatic impact wrench can be fitted at the end of the arm just as easily as the grippers.

Grippers are made in a variety of sizes and shapes. See Figure 2–12. They are designed for special applications by the manufacturers of robotic equipment. The simplest type of gripper is a motion-producing device that is joined by two fingers. The fingers open and close to grasp an object. In most instances, only the finger mounts are purchased and the designing and building of the fingers are done in the local machine shop to fit the job being done. See Figure 2–13. Variations in design account for much of the design time spent on robots.

Various types of grippers are made to fit the job being done by the robot. For instance, it is possible to have a donut-shaped inflatable piece of rubber that slips over the neck of a bottle, is inflated, and the bottle then moved. When it is time to release the bottle the donut is deflated, the bottle slides out, and the robot arm moves upward and onto the next bottle that needs to be lifted onto the line. In some cases rubber suction cups are used to pick up materials. See Figure 2–12, 9 and 10. A vacuum is applied to pick up the object and a blast of air is added to cause the object to drop.

1. Machine tools   6. Woodworking machines
2. Stamping presses   7. Welding fixtures
3. Moulding machines 8. Extrusion billets
4. Die cast machines   9. Transfer equipment
5. Forging presses   10. Glass handling

FIGURE 2–12   Different types of grippers made for the end of the manipulator arm (courtesy of I.S.I. MANUFACTURING, INC.)

FIGURE 2-13   Three-fingered gripper, air operated (courtesy of COMPACT AIR PRODUCTS, INC.)

☐ ALL TOGETHER IT BECOMES A MANIPULATOR

The manipulator is really a combination shoulder, arm, wrist, and hand. See Figure 2-14. The grippers are the hand. This combination makes it

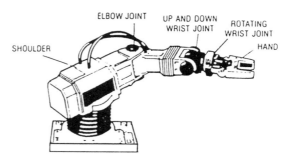

FIGURE 2-14   The parts of an industrial robot used to pick and place (courtesy of RADIO SHACK)

possible for a robot to reach for an object, pick it up, carry it, and put it down where desired.

Work Envelope

The *work envelope* is also referred to as a sphere of influence or work area since it is not always spherical in shape. This is the space a robot occupies when it swings around and up and down to do the work for which it is designed. See Figure 2-15.

Articulation

Three *articulations* are needed for a robot to move its arm, wrist, and hand to any place within its work envelope. The articulations are: (1) Extend and retract the arm; this can be simply moving the arm out and in. (2) Swing or rotate the arm; this is moving the arm left and right. (3) Elevate the arm; this is nothing more than lifting or depressing (lowering) the arm.

Wrist Motions

The three types of motion for a robot's wrist are: (1) bending or rolling forward and backward; (2) yawing (spinning) from right to left or left to right; and (3) swiveling, which is nothing more than rolling down to the right or left. See Figure 2-16.

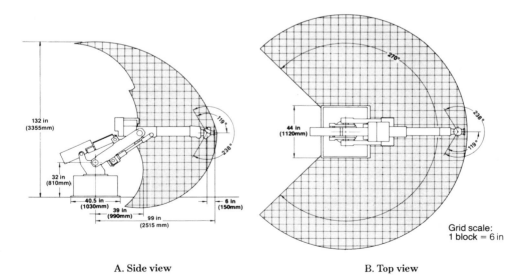

A. Side view   B. Top view

FIGURE 2-15   Tear-shaped work envelope

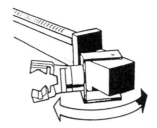

FIGURE 2-16   Wrist action known as yaw (courtesy of CINCIN-
NATI MILACRON)

Degrees of Freedom

The robot has six *degrees of freedom* if it can move the wrist three ways
and the arm three ways. See Figure 2–9. This is very limited when you
compare it with human shoulder, arm, and wrist motions. Humans have
forty-two degrees of freedom. As you can see, the robot arm needs improvement if it is to be as versatile as the human arm. There is some
question as to whether the human arm and wrist should be emulated.
However, such emulation would call for some rather involved engineering to get the job done right. At the present time it is possible for the
robot hand to do what the human hand can do: grip, push, pull, grasp,
and release.

☐ ROBOT MOTION CAPABILITIES

Robots have four basic motion capabilities: (1) linear motion, (2) extension motion, (3) rotating motion, and (4) twisting motion. These four motion capabilities are referred to as the *LERT* classification system. (*L*
stands for *l*inear motion; *E* stands for *e*xtensional motion; *R* stands for
*r*otational motion, and *T* stands for *t*wisting motion.) The superscript
used with the classification system indicates the number of times the
robot is capable of a particular motion. Keep in mind that most robots
are mounted to the floor on a base. However, it is possible for them to be
mounted to the ceiling or onto a mobile platform. Each axis is listed in
order as it is mounted to the first component or base. For instance, L^3
indicates there are three linear motions. If you use the R^2L^3, you have
two rotational motions and three linear motions.

☐ COORDINATES

The manipulator arm geometry refers to the movement of the robot
arm. There are four systems of classification for robot axis movement:

articulate, Cartesian, cylindrical, and polar. Each of these *coordinates* describes the arm movement through space. Various axis movements are the result of design characteristics.

Cartesian Coordinates

The simplest and most easily understood coordinate system is the Cartesian. *Cartesian coordinates* are used to describe the X, Y, and Z axes or planes. The reference point for these planes is the intersection of all three. See Figure 2–17. The centerline of the robot is the reference point for all three axes. Note the location of the center point or origin point.

The manipulator uses the X, Y, and Z planes to reach its *target*, so these planes become the operational axes of the manipulator arm. Note how the movement along the Z axis is in a linear motion with an up-and-down movement from point a to point b. See Figure 2–18. The Y axis is also a linear motion but with an in-and-out movement. This Y axis provides a reaching type of movement of the arm from point b to point c. The X axis is the side-to-side motion of the manipulator. That means the side-to-side motion produces a movement that causes the whole manipulator to rotate about its base.

A rectangular-shaped work envelope is produced by the Cartesian coordinate robot. This rectangular working area is caused by the limitations that the X, Y, Z axes produce. It is not possible to reach areas out-

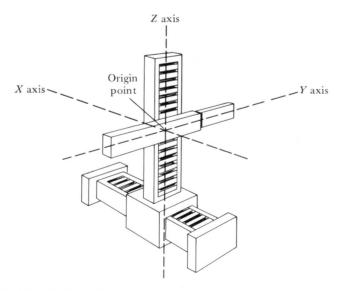

FIGURE 2–17   Cartesian coordinates or X, Y, and Z axes (from Malcolm *Robotics*)

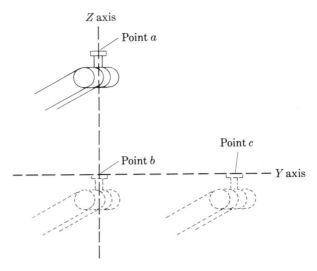

FIGURE 2-18 Manipulator arm movements for a Cartesian coordinate system (from Malcolm *Robotics*)

side the rectangular work envelope with a Cartesian coordinate system. It is impossible for the arm to move from point a to point c directly. The movement is limited to the X, Y, and Z directions.

The Cartesian coordinate system robot is one of the simplest in its operation. It can be used to load and unload and do point-to-point operations. This type of robot can be further classified as a low technology type. Cartesian in of itself does not make it low technology. Most air operated pick-and-place robots, being low technology, are of the Cartesian design. See Figure 2–19.

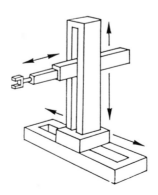

FIGURE 2-19 Cartesian coordinates

Cylindrical Coordinates

The cylindrical coordinate system also uses three axes; however, the names for the axes are different. See Figure 2–20. Cylindrical coordinates are labeled theta (θ) and describe the rotational axes. See Figure 2–21. The R axis is the reach or in-and-out axis. The Z axis indicates the up-and-down movement. This type of movement results in tracing out a cylindrical shape when its work envelope is examined. Keep in mind that the rotation of the base is described by the Greek letter theta (θ). The cylindrical robot can rotate up to 300°. The extra 60° are used for a safety zone around the robot. This safety zone is called the *dead zone*. See Figure 2–22. This type of robot can reach from 19 inches to 59 inches depending on the design and task to be performed. The reach and other features of the robot are usually described in metric measure-

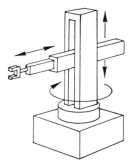

FIGURE 2–20   Cylindrical coordinates

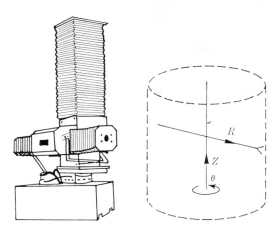

FIGURE 2–21   Cylindrical robots and the resulting cylinder traced by the arm's movements (courtesy of PRAB ROBOTS, INC.)

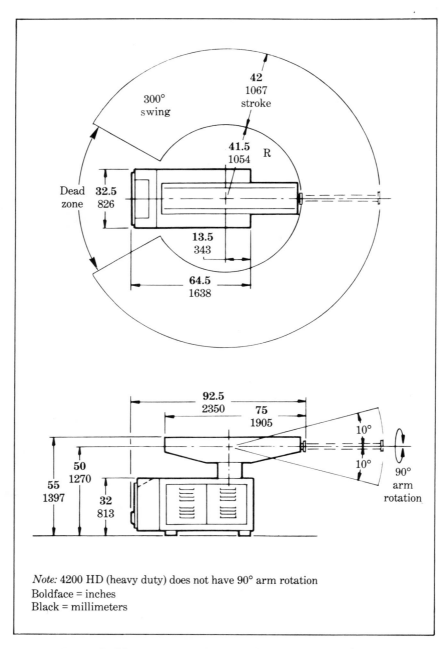

FIGURE 2–22   Location of the dead zone (courtesy of PRAB RO-
BOTS, INC.)

ments. Thus the robot has a capability of reaching 500 to 1,500 millimeters (mm). An inch is equal to 25.4 mm.

Z axis movement is the up-and-down motion. Most robots can travel within 100 to 1,100 mm (about 4 to 43 inches). The amount of movement is determined by the task to be performed.

Polar Coordinates

The polar coordinate system is slightly different, but it too has three axes of operation. The polar coordinate system describes a *spherical* movement pattern.

The three axes of this system are theta (θ), R or reach axis, and beta (β). As you can see from examining Figure 2–23, the polar system and the cylindrical system basically have the same axes. Theta and R axes are the same in both types of robots. However, the beta axis allows the entire arm of the robot to move in an up-and-down motion.

Articulate Coordinates

The articulate coordinate system is descriptive of a *jointed* system. The theta motion or rotation around the base is already familiar to you. See Figure 2–24. It, too, is used in this system. However, there is a slight change in the designation and motion of the other two axes. The W axis is the upper arm or shoulder and the U axis is the elbow motion. Two axes in this type of robot move or bend. The W axis represents the shoulder rotation, whereas the U axis describes the motion of the elbow and its rotation. This produces a greater degree of flexibility in the robot, making it ideal for industrial applications.

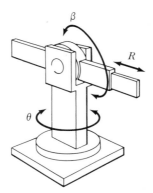

FIGURE 2–23   The three axes of a polar coordinate robot (from Malcolm *Robotics*)

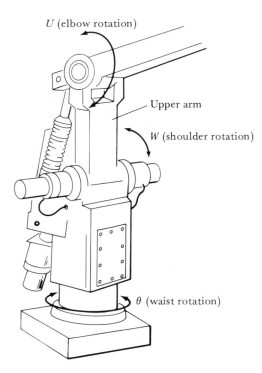

FIGURE 2-24   The three axes of an articulate coordinate robot (from Malcolm *Robotics*)

☐ WRIST ACTION

Coordinate systems are also used to describe the motion possibilities of the wrist. The manipulator arm is limited in its ability to do work without some type of *end effector* to act as a hand. To make this hand work, it is necessary to have a wrist action to cause it to approach the abilities of the human hand to do work.

Four coordinate systems are used to describe this wrist action. You can see the possibilities of the wrist in Figure 2-25. By adding the wrist, it is possible to have a robot with six axes of motion. This added dimension increases the flexibility of the robot and its possible application list. The axes that the wrist adds to the robot are identified as the pitch axis, the yaw axis, and the roll axis.

The pitch axis is usually limited to 270° of movement. It moves up and down. This 270° can be moved gradually or degree by degree, depending on the task to be performed. See Figure 2-26.

The yaw axis describes the side-to-side movement of the wrist. The movement range is 90° to 270°.

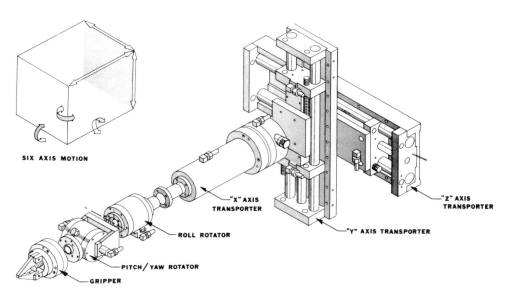

FIGURE 2–25 Wrist action (Note how the movements are made by rotators.) (courtesy of MACK CORPORATION)

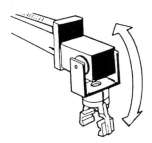

FIGURE 2–26 The wrist action known as pitch (courtesy of CINCINNATI MILACRON)

The roll axis refers to the movement of the end of the wrist. It is possible to obtain 360° of rotation with an end effector.

With these three additional axes of rotation, it is possible to have the robot do things that humans cannot. This is especially true in painting automobiles on the assembly line. It makes possible the painting of the inside of the car body at the same time the outside is sprayed, which

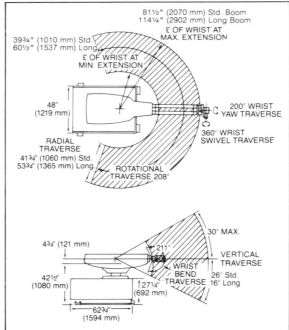

FIGURE 2-27 Spherical robot and its work envelope (courtesy of UNIMATION INCORPORATED, a Westinghouse Company)

results in better protection of the body metal with decreased possibility of rust later.

☐ WORK ENVELOPES

The Cartesian coordinate robot has a rectangular-shaped work envelope. This envelope is very important to anyone who works on or near robots. This is the area in which the robot can move. Since the robot cannot see, it is possible for its arm to hit a person or equipment that is left in this area. The cylindrical coordinate robot has a cylindrical work envelope, which means you also have to be careful of what is above the robot as well as around it.

The polar coordinate robot (see Figure 2–27) has a spherical-shaped work envelope and the articulate coordinate work envelope is tear-shaped. Keep in mind that all points that can be programmed within the reach of the robot are part of the work envelope.

Look at Figure 2–28 to obtain a better idea as to the space occupied by various types of robots in their normal operational mode.

☐ MOVING THE MANIPULATOR

In order for the manipulator to do work it must have some *power source*. Industrial robots work on 220-volt and 440-volt systems available within the plant or factory. Entertainment and smaller robots may work with batteries that need recharging occasionally. All robots use electricity to give them their basic power. However, there are other drive methods used to produce motion and do work. They are hydraulic and pneumatic. There are, then, three ways to drive the manipulator and do the task the robot was designed to do. Technically, there are four. Mechanically driven robots can also be counted as a drive method.

Pneumatic Drive

About one-third of industrial robots use pneumatic power to drive the axes. If some of the pneumatic powered pick-and-place units are not considered robots, then the percentage may be less. *Pneumatic power* is generated by compressed air. Air is pressurized or compressed within the plant and fed through pipes to different locations. When the pressurized air is fed to a cylinder attached to some part of the robot, it is converted into motion. The air pressure causes desired movement of the arm or the whole robot. See Figure 2–29.

Pneumatic drive systems have some advantages, among them cost. They are the least costly of the three drive sources. However, pneumatic drive systems do have some limitations. They can develop only

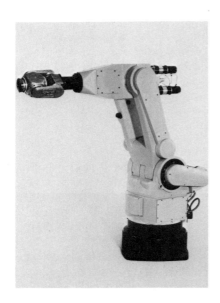

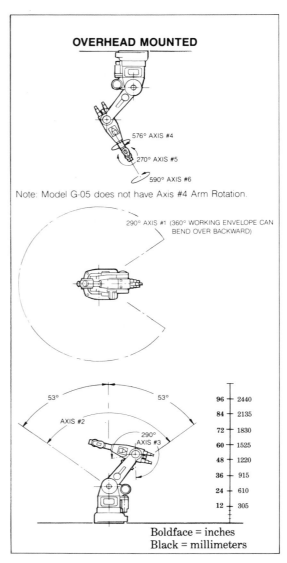

OVERHEAD MOUNTED

576° AXIS #4

270° AXIS #5

590° AXIS #6

Note: Model G-05 does not have Axis #4 Arm Rotation.

290° AXIS #1 (360° WORKING ENVELOPE CAN BEND OVER BACKWARD)

53°   53°

AXIS #2

290° AXIS #3

96	2440
84	2135
72	1830
60	1525
48	1220
36	915
24	610
12	305

Boldface = inches
Black = millimeters

FIGURE 2-28 A robot at work (courtesy of PRAB ROBOTS, INC.)

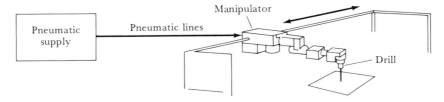

FIGURE 2-29 Pneumatic-operated robot system (from Malcolm *Robotics*)

enough torque to pick up about 6.5 to 10 pounds (3 to 4.5 kilograms). Pneumatic drive systems are used on robots that do simple assembly operations, die-casting operations, materials handling jobs, and machine loading and unloading. Their greatest limitation is their inability to move large payloads. Larger payloads require the electric or hydraulic drives. See Appendix I.

Hydraulic Drive

About 45 percent of the industrial robots in use today are driven by hydraulic systems. Hydraulic fluid is placed under pressure by a pump operated by an electric motor. The pressurized fluid is used to move the axis of the robot by applying this pressure to a cylinder, hydraulic motor, or rotary actuator. The cylinder is extended by the pressure of the fluid. The direction of the fluid flow in the cylinder determines if it extends or retracts. By moving the cylinder in and out it is possible to move the axis of the robot.

Hydraulic drive systems are capable of lifting 50 to 300 pounds (22.7 to 135.9 kilograms). They are usually referred to as medium technology robot systems, but some hydraulic systems can be considered high technology. Low, medium, or high technology classifications depend on the type of control not the power source.

Hydraulic drive systems can be used to load and unload up to their lifting ability. They are also used in arc welding, spot welding, die casting, spray painting, and press loading. One of their major uses is in spray painting because they are safer to operate in a volatile atmosphere than are electric powered robots. Figure 2–30 shows the simplicity of the hookup for a hydraulic drive system.

Some disadvantages are associated with the use of hydraulic systems. The fluid is under pressure and is difficult to contain. Cost is another factor to consider. Depending on the size of the unit being considered, cost can be of importance. Generally speaking, hydraulic systems are the most expensive of the three drive systems mentioned here. However, they are employed where a lot of torque is required to lift heavy loads.

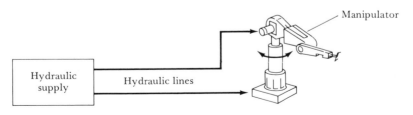

FIGURE 2–30 Hydraulic-operated robot system (from Malcolm *Robotics*)

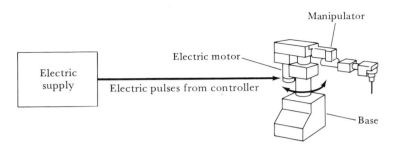

FIGURE 2–31   Electric-operated robot system (from Malcolm *Robotics*)

Electric Drive

Electric drive systems use electric motors for their power source. The motors may be operated by direct current (DC) or alternating current (AC). Torque is developed by gearing the speed of the motor down to the required motion or movement of the arm. See Figure 2–31.

Some advantages of electric drive systems are their ability to allow smooth start-up of payloads and their smooth deceleration and stopping. Operation costs are minimal in terms of electrical power used and maintenance of the system motors. Electric robots have a higher repeatable accuracy than do hydraulic robots.

Electric drive robots cannot handle as heavy a payload as hydraulic drive systems. The payload for an electric drive is between 6.6 and 176 pounds (3–80 kilograms). See Appendix I. However, electric drive systems are very versatile in their operation.

Electric drive systems are used in arc welding, spot welding, machine loading and unloading, materials handling, and deburring as well as in assembly operations. Development of explosion-proof motors will make it possible to use them safely in spray painting atmospheres.

Electric drive systems are used in high technology robots.

☐ SUMMARY

Robots are made up of a number of subsystems. There are a number of methods used to classify robots. The classification system used in this book is according to the end purpose of the robot.

Industrial robots have arms with grippers attached. The grippers are fingerlike and can grip or pick up various objects. They are used to pick and place. Robots can be computerized and operate without human supervision.

Laboratory robots take many shapes and do many things. They have microcomputer brains, new multijointed arms, and advanced vision or tactile senses. Some have hand-eye coordination.

The explorer robot is used to probe outerspace and to explore caves, dive underwater, and explore areas where no human can exist.

Most hobbyist robots are mobile. They are still experimental and have been part of an effort to develop a housekeeper that resembles the human form.

Classroom robots have limited application today. However, remarkable things are planned for them in the near future.

Entertainment robots are just beginning to be developed and made available to entertain people and act as roving advertisements.

The manipulator is one of three basic components of the robot. The other two are the controller and the power supply. The manipulator can be classified by four coordinate systems used to describe the arm movement.

The base of the robot is its anchor point. The base may be either rigid or mobile.

Some type of arm is found on most industrial robots. It may be jointed and resemble a human arm or it may be a slide-in/slide-out type used to grasp something and bring it back closer to the robot.

The wrist is attached to the jointed arm and can be designed with a wide range of motions.

The gripper is located at the end of the wrist and is used to hold whatever the robot is to manipulate.

The manipulator is really a combination shoulder, arm, wrist, and hand. The gripper represents the hand.

Work envelope is also referred to as a sphere of influence or work area for the robot.

Articulations are (1) extend and retract the arm, (2) swing or rotate the arm, and (3) elevate the arm.

The robot has six degrees of freedom if it can move the wrist three ways and the arm three ways. Human workers have forty-two degrees of freedom.

The four basic motion capabilities of robots are linear motion, rotating motion, twisting motion, and extensional motion. These four motions are the basis of the LERT classification system.

The manipulator arm geometry refers to the movement of the robot arm. There are four systems of classification for robot axis movement: articulate, Cartesian, cylindrical, and polar. Each describes the movement of the arm through space and within its work envelope. The manipulator uses the X, Y, and Z planes to reach its target. However there are also the theta (θ), beta (β), W, and U

axes to be considered in the operation of a robot. Coordinate systems are also used to describe the motion possibilities of the wrist. The manipulator arm is limited in its ability to do work without some type of end effector to act as a hand. The axes that the wrists adds to the robot are identified as the pitch axis, yaw axis, and roll axis.

Cartesian coordinate robots have a rectangular-shaped work envelope. The polar coordinate robot has a spherical-shaped work envelope, and the articulate coordinate type is tear-shaped. The work envelope for the cylindrical coordinate robot is cylindrical in shape.

Drive systems for robots are classified as pneumatic, hydraulic, and electric. Each type has its applications due to physical limitations of the method used to do the work. Each type of drive has its advantages and disadvantages.

☐ KEY TERMS

articulations   the ability of a robot to extend and retract, swing or rotate, and elevate its arm

Cartesian coordinates   simplest of the coordinates because it refers to up-and-down, back-and-forth, and in-and-out movement of a robot arm

controller   supplies the direction to the robot

coordinates   points and planes in reference to the movement capability of a robot arm

degrees of freedom   the number of axes found on a robot

dead zone   safety zone where the arm does not move during normal operation

end effector   device mounted on the end of the manipulator or robot arm to physically do the work

grippers   located on the end of the manipulator arm and used to pick up things

LERT   classification system for robots based on four basic motion capabilities: *l*inear, *e*xtensional, *r*otational, *t*wisting

manipulator   one of three basic parts of a robot

microprocessor   a device, a semiconductor chip, used to provide the brain for robot control

pneumatic power   using air pressure for driving the manipulator

power source   supplies power to the robot

target   the point at which the arm is expected to reach for picking up an object

work envelope   that space in which the robot arm moves during its normal work cycle

☐ QUESTIONS

1. List six types of robots according to end purpose.
2. What is a manipulator?
3. What is the anchor point of the robot called?
4. How many axes of motion does the wrist have?
5. What is a gripper?
6. What is an end effector?
7. What are the four basic motion capabilities of a robot?
8. What does arm geometry mean?
9. What is the target in robot terminology?
10. What is a Cartesian coordinate?
11. List the four classifications of robots according to the coordinate system.
12. What is a work envelope?
13. What type of coordinate system produces a tear-shaped work envelope?
14. What does yaw mean?
15. What is a roll axis?
16. What are the three uses for the electric drive type of robot?
17. What is the advantage of the pneumatic method of moving a robot?
18. What is the disadvantage of the electric drive robot?
19. How much can the hydraulic drive robot lift?
20. How much can the electric drive robot lift?

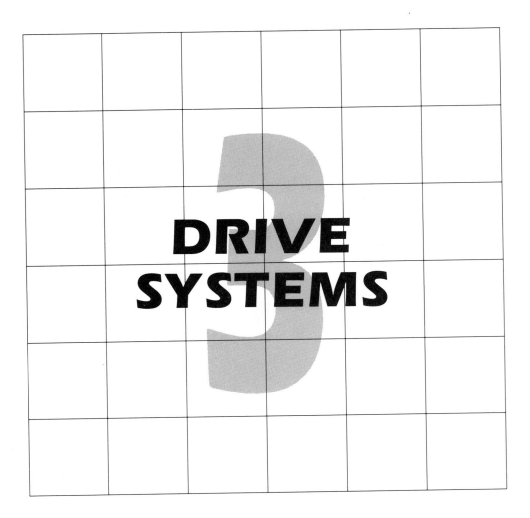

DRIVE SYSTEMS

R obots need some type of power to cause them to function. In order to make the arm or any other part of the system move, it is necessary to develop the power in a usable form and in some type of readily available unit.

Three ways used to power a robot and its manipulators have already been mentioned. They are hydraulic systems, pneumatic systems, or electricity. Of course, electricity is used to make the pneumatic and hydraulic systems operate. However, there is also an all-electric type of drive available in the form of electric motors being directly attached to the manipulator or moving portion of the robot.

☐ HYDRAULICS

Of the three types of drive systems for robots, the hydraulic system is capable of picking up or moving the heaviest loads, which is why it is used. See Figure 3–1. It is ideally suited for spray painting where electric systems can be hazardous.

The automobile uses a hydraulic system for braking. As the brake pedal is depressed, it puts pressure on a reservoir of liquid that is moved under pressure to the brake cylinders. The brake cylinders then apply pressure to the pads that make contact with the rotor that is attached to the front wheel. By applying the pressure to the pads in varying amounts, forward motion of the car is either stopped or slowed, as desired. The hydraulic system used for the robot is similar to the braking system of an automobile.

Hydraulics is the Greek word for water. However, oil, not water, is used in robot drive systems. (Some water-based hydraulic fluids are used in foundry and forging operations.) It is this oil that presents the pressure to the correct place at the right time in order to move the manipulator or grippers. Hydraulic pressure in a hydraulic system applies force to a confined liquid. The more force applied to the oil, the greater the pressure on the liquid in the container.

Pressure

The force applied to a given area is called *pressure*. Pressure is measured in *pounds per square inch* (in metrics the unit of measurement is kilopascals), which can be appreviated as lb/in^2 or psi.

The SI metric unit for pressure is the *pascal*. The pascal is one newton per square meter (newton is the unit of force). Because the pascal is such a small unit, kilopascals (kPa) and megapascals (MPa) are used. For example, a typical automobile tire pressure equals 190 kPa.

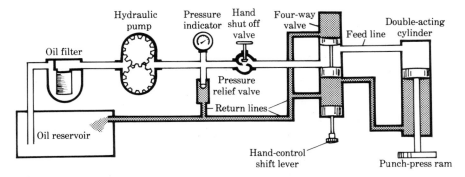

FIGURE 3–1   Hydraulic system

Typical industrial air pressure lines equal 700 kPa. One lb/in^2 is equal to 6.896 kPa, and 1 kPa is equal to 1.45 lb/in^2.

The pressure applied to a hydraulic system is in terms of how much push or force is applied to a container of oil. See Figure 3–2. There are several ways to develop the pressure needed in a robot. The pressure developed in a hydraulic circuit is in terms of how much push or force is applied per unit area. Pressure results because of a load on the output of a hydraulic circuit or because of some type of resistance to flow. A pump is used to create fluid flow through the piping to the point where it causes movement of the gripper or manipulator of the robot. A weight can also be used to generate pressure.

One of the disadvantages of using the hydraulic system is its inherent problem with leaks. While fluid is under pressure, it has a tendency to seek the weakest point in the system and then run out. This means that each joint along the piping must be able to withstand high pressure without leaking. It also means that the moving part of the robot, the manipulator, must be able to contain the fluid while it is under pressure. In most cases the O rings are not able to contain the fluid and leaks occur. O rings are very important in sealing the moving end of the manipulator or terminating end of the fluid line. This is one of the maintenance problems associated with robots used for picking up heavy objects. The hydraulic system is used to pick up the weight, and then it has to be able to maintain the integrity of the system to keep from losing too much fluid. The fluid is slippery when spilled on the floor and must be contained to prevent accidents by humans working in the area.

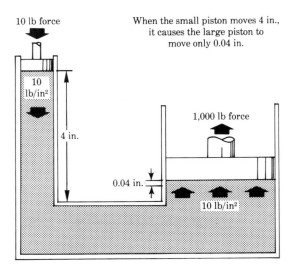

FIGURE 3–2 Hydraulic pressure changes (The 10-pound force can cause 1,000 pounds to be lifted, but notice the distance of each.)

Because of this pressure build-up in a hydraulic system, it is necessary to prevent its continuing to increase beyond the capacity of the pipes and containers. That is where a relief valve becomes important. A relief valve is used as an outlet when the pressure in the system rises beyond the point where the system can handle it safely. The valve is closed as long as the pressure is below its design value. Once the pressure reaches the point of design for the relief valve, it opens. Then the excess fluid must be returned to the reservoir or tank. This means a relief valve and the return of the fluid to the tank must be part of the system.

Hydraulic systems need filters to keep the fluid clean. Even small contaminants can cause wear quickly. The filter has to take out contaminants that measure only microns in size. The filters have to be changed regularly to keep the fluid and system in operating condition.

Pressures of several thousand psi are not uncommon in the operation of a robot. This means that a leak as small as a pin hole can be dangerous. If you put your hand near a leak, it is possible that the fluid being emitted through the hole can cut off your hand before you feel it happen. However, this will not occur at pressures under 2,000 psi. High pressures are very dangerous.

Additional volume for the operation of the robot may be obtained by pressurizing the surge tank with a gas. An accumulator can only be charged to the main system pressure. In some cases the robot will require a larger volume of fluid for rapid motion than the pump alone can provide. In such cases a surge tank on the high pressure side of the line is charged with a gas in a flexible container. This flexible container is inside the tank and expands. High pressure of the fluid compresses the gas when there is excess pressure available. Then, when the system needs the fluid for the rapid motion, the pressure drops slightly and the gas in the tank expands to force the extra fluid into the system.

Hydraulic motors mounted on a manipulator are operated by signals from a transducer either to open the valve for a given period of time or to keep it closed. The control system, discussed later, is necessary to make the robot do what it is supposed to do when it is supposed to do it. Hydraulic motors and cylinders are compact and generate high levels of force and power. They make it possible to obtain exact movements very quickly. See Figure 3–3.

☐ PUMPS

The hydraulic system must have a pump in order to operate. The pump converts mechanical energy, usually supplied by an electric motor, to hydraulic energy. Pumps are used to push the hydraulic fluid through the system.

There are two basic types of hydraulic pumps: hydrodynamic and hydrostatic. The hydrodynamic pump is a low resistance pump and is

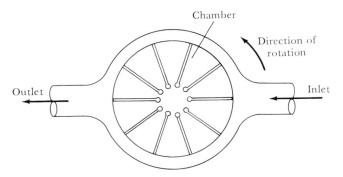

FIGURE 3-3 Hydraulic motor (Pressure on hydraulic fluid causes the motor to turn in a counterclockwise direction. Reversing the direction of fluid flow reverses the direction of motor rotation.) (from Malcolm *Robotics*)

not found in robot systems. Therefore, a closer look at the hydrostatic pump is in order.

Hydrostatic pumps are further classified into gear pumps and vane pumps. These pumps are called on to deliver a constant flow of fluid to the manipulator. The gear pump produces its pumping action by creating a partial vacuum when the gear teeth near the inlet unmesh, causing fluid to flow into the pump to fill the vacuum. As the gears move, they cause the fluid or oil to be moved around the outside of the gears to the outlet hole at the top. The meshing of the gears at the outlet end create a pushing action that forces the oil out the hole and into the system.

The vane pump gets its name from its design, as does the gear pump. The vanes cause the fluid to move from a large volume area to a small volume area. By pushing the fluid into a smaller area, the pressure on the fluid increases since the fluid itself is not compressible. See Figure 3-4. A vane pump can be used to supply low-to-medium pressure ranges, capacity, and speed ranges. They are used to supply the manipulator with the energy to lift large loads. Pressures can reach as high as 2,000 psi with about 25 gallons of fluid per minute moving in the system.

Piston-type rotary devices are also used to increase the pressure in a hydraulic system. The pistons retract to take in a large volume of fluid and then extend to push the fluid into the high pressure output port. There are usually seven or nine cylinders. These vary with the different configurations of machines. The in-line piston pumps are good for robot applications. They have a very high capacity, and their speed ranges are from medium to very high. Pressures can get up over 5,000 psi. See Figure 3-5.

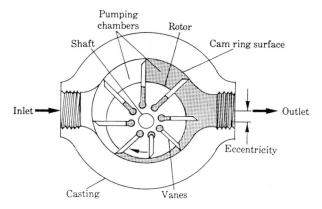

FIGURE 3-4   Vane-type pump/motor (Note that this hydraulic pump can also be used as a motor when driven by the hydraulic fluid under pressure.)

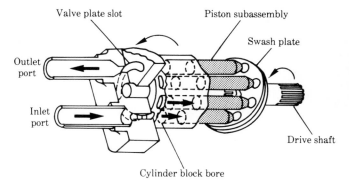

FIGURE 3-5   Piston-type rotary pump

☐ PNEUMATICS

Pneu simply means *air* in Latin. Pneumatics is that part of physics that works with air and gases. Robots use pneumatics to control the grippers and in some cases the manipulator arm. Therefore, in the field of robotics, compressed air is the medium used.

   Pneumatic systems are used in industry to power hand tools and to lift and clamp products during machining operations. They use a compressor with a tank to store the compressed air. The compressor is driven by an internal combustion engine or an electric motor. Once the air is compressed, it is filtered to remove all dirt and moisture. An air filter is used to remove the contaminants, and a condensation trap and drain are used to remove the moisture. In some cases a mist of oil is added to aid in the lubrication of the parts being serviced by the air supply. See Figure 3–6.

In order to maintain a constant pressure, the motor kicks on when the pressure in the storage tank drops to a predetermined level. The motor and compressor then build up the pressure and turn off until needed again. An advantage that the pneumatic system has over the hydraulic system is that it can exhaust compressed air into the atmosphere, whereas the hydraulic system has to have a return system for containing the fluid for reuse.

Pneumatic systems contain a motor-driven compressor, a storage tank, and lines to carry the air from the compressor or storage tank to the using device. Along the way the system has several methods of control. There is the hand shutoff valve and pressure-relief valve to control the air. Air flow or pressure can be regulated by a regulator and a three-way valve. Notice the similarity of the pneumatic and hydraulic systems. However, the pneumatic system simply exhausts the air into the atmosphere when it is finished using the pressure to operate the device. Silencers keep down the noise of the exhausted air. The hydraulic system has to have a return line from the user to the storage tank. This makes the hydraulic system more expensive to install and operate.

The pneumatic cylinder is the load device designed to use air pressure to do work. This part changes the mechanical energy of air into linear motion that drives a press ram. Pneumatic load devices can also be used to produce rotary motion.

About the only problem in maintaining a pneumatic system is keeping the air supply free of moisture and dirt. The lines have to be kept clean and dry.

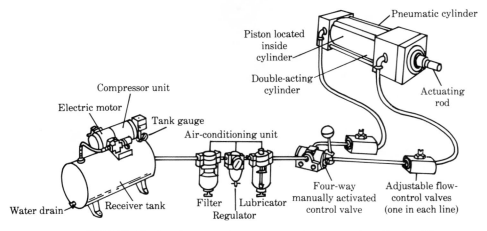

FIGURE 3–6   Pneumatic system and pneumatic cylinder (The piston and actuating rod are pushed forward by air pressure.)

☐ ELECTRIC MOTORS

Electric motors are very useful in robotics. They are easily controlled with a computer or microprocessor, and they can be reversed easily. Another reason for using electric motors in robots is the ease with which the torque and speed can be controlled.

Electric motors can be used on direct current (DC) or alternating current (AC). However, each type has its own limitations and uses. It is not the purpose here to cover the operation of various types of electric motors. A brief description of the operation of the types used for robots is given.

DC Motors

The DC motor has the series, shunt, and compound configurations. See Figure 3-7. Each type has advantages and disadvantages. The series, for instance, can be used where there is a need for lots of torque to lift or move something. However, it cannot be used for devices that are belt-driven or where a constant speed is needed. The shunt does not have the starting torque but does have the constant speed characteristic. The compound has some of the qualities of both the series and shunt but has other limitations.

Most DC motors have brushes. They also have commutators that consist of small pieces of copper separated by a mica insulator. A maintenance problem can result because of brush wear and the arc that occurs between the brushes and the commutator as the motor is loaded.

Speed control of the DC motor can be accomplished by regulating its voltage or current or both. Electronic controls have been designed to aid in the speed control of DC motors. They are easily reversed by reversing the polarity of the electric current supplied to their windings. Variable resistors can also be used to control the speed of DC motors. The resistor is inserted in series with the field windings and adjusted to increase or decrease the voltage available to the motor.

PM Motors

The permanent magnet (PM) motor is used to drive small toys and to drive electric seats and electric windows in automobiles. They have other uses as well. They are not often used in robots except in toy applications. They consist of a permanent magnet and a wound rotor or armature. The direct current is fed to the armature through brushes and commutator segments, and it sets up a magnetic field that is attracted to the permanent magnet's field. Once the unlike poles have been attracted and the armature rotates toward the fixed permanent magnet, the commutator segments that furnished the DC power to the armature move over to two other segments and cause a different pole to be ener-

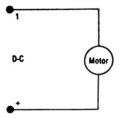

PERMANENT-MAGNET, BRUSH-LESS DC, PRINTED CIRCUIT, SHELL-TYPE ARMATURE. To reverse: transpose motor leads.

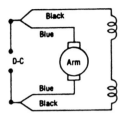

SHUNT WOUND. To reverse: transpose blue or black leads.

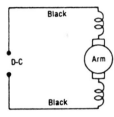

SERIES WOUND (2 LEAD). Non-reversible.

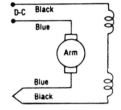

SERIES WOUND (4 LEAD). To reverse: transpose blue leads.

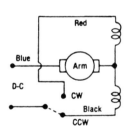

SERIES WOUND (SPLIT FIELD). To reverse: connect other field lead to line.

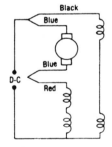

COMPOUND WOUND (5-WIRE REVERSIBLE). To reverse: transpose blue leads.

FIGURE 3-7 Wiring diagrams for DC motors (The loops indicate field coils, and *arm* indicates the armature or rotor.)

gized and magnetized. This pole will then be attracted to the permanent magnet. But as it approaches the permanent magnet, the energy in the coil of wire making up this segment will be removed by the commutator segment moving on to make contact with the brushes on the next two segments. This continues since the commutator is nothing more than a switching device that determines which coils are energized and when.

DC Brushless Motors

The newer DC *brushless motors* use transistors or "Hall effect" magnetic devices to reverse the current through the field coils. Whereas the commutator does the switching of current in the armature of regular DC motors, the DC brushless motor uses electronics for the switching operation. There are two types of brushless motors: the Hall effect and the split-phase, permanent magnet.

The split-phase, permanent magnet motor has center-tapped windings. See Figure 3–8. Two transistors are used to set up an oscillator circuit to supply power to the windings. Resistance, capacitance, and inductance of the motor windings determine the frequency at which the oscillator operates. This, in turn, determines the operation of the motor.

The Hall-effect motor uses that quality of a transistor to react to the presence of a magnetic field to do its switching. When a magnetic field passes near a transistor, the semiconductor's resistance decreases. This decrease in resistance causes the current in the circuit to increase. The increased current flow is fed to the electronic switch that controls the current being fed to the field winding of the motor. See Figure 3–9.

The permanent magnets of the brushless motors are mounted on the armature shaft rather than being mounted as field magnets. The field coil is wound.

The advantage of using a brushless motor is its long life. The

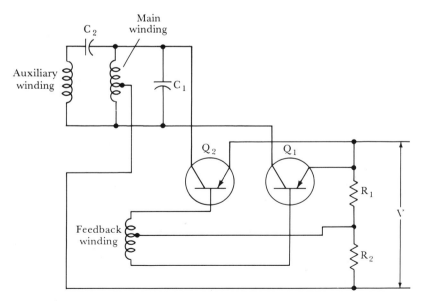

FIGURE 3–8   Split-phase, permanent magnet DC brushless motor (from Malcolm *Robotics*)

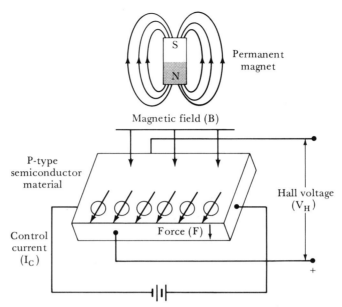

A. Downward direction of current to generator Hall voltage and magnetic field

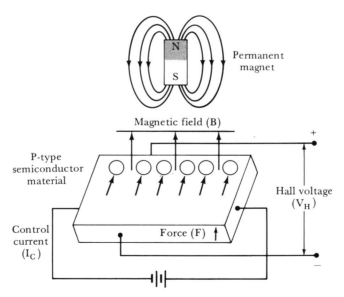

B. Reversed direction of magnetic field

FIGURE 3-9 Hall effect used for controlling motor speed (from Malcolm *Robotics*)

elimination of the brushes and commutator makes it practically a maintenance-free motor. The main disadvantage of the brushless motor is the low torque developed by the split-phase, permanent magnet type.

Stepper Motors

Stepper motors are used with educational robots that demonstrate the basic operation of robots. They are not presently used in industrial applications. They have the disadvantage of slipping if overloaded. This means the error created by the slippage can go undetected and ruin whatever is being machined or processed. See Figure 3–10A.

The stepper motor is used primarily to change electrical pulses into rotary motion that can be used to produce mechanical movement, which is why they are so well mated to computers. The computer then generates pulses needed to operate the stepper motor. Operation of the bipolar stepper motor is accomplished in a four-step switching sequence. Any of the four combinations of switches 1 or 2 will produce an appropriate rotor position location. See Figure 3–10B. After the four switch combinations have been achieved, the switching cycle repeats itself. Each switching combination causes the motor to move one step.

Some stepper motors use eight switching combinations to achieve "half stepping." During this type of operation, the motor shaft moves half of its normal step angle for each input pulse applied to the stator. This allows for a very precise and controlled movement. See Figure 3–10C and the eight stator windings.

☐ AC MOTORS

AC motors are used to operate the air compressors and the materials handling equipment in an industrial plant. They furnish the energy to move materials and equipment. Direct current is used to control the movement of robots, in some cases because it is easier to control the rotation speed of a DC motor than it is an AC motor. DC motors also better lend themselves to control by computers than do AC motors. However, recent developments make it possible for AC motors to be more accurately controlled by computers.

The two types of AC motors most frequently used on robots are wound-rotor induction and squirrel-cage types. They operate on 120, 240, and 440 volts according to the voltage available in the plant where the robot is located.

Induction Motor

This classification includes a large variety of motors. However, we will limit our discussion to the wire-wound rotor induction motor. This type

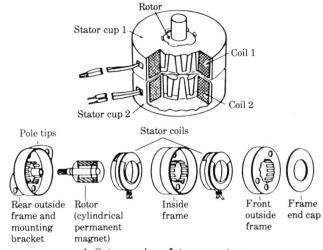

Rotor

Stator cup 1

Coil 1

Coil 2

Stator cup 2

Pole tips

Stator coils

Rear outside frame and mounting bracket

Rotor (cylindrical permanent magnet)

Inside frame

Front outside frame

Frame end cap

A. Cutaway view of stepper motor

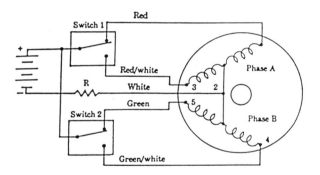

Red

Switch 1

+

R

Switch 2

Red/white

White

Green

Green/white

Phase A

3 2

5

Phase B

Switching Sequence*		
Step	Switch #1	Switch #2
1	1	5
2	1	4
3	3	4
4	3	5
1	1	5

*To reverse direction, read chart up from bottom.

B. Wiring diagram and switch combinations

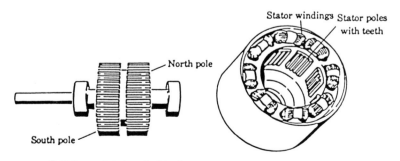

Stator windings Stator poles with teeth

North pole

South pole

C. Rotor and stator of a bipolar, permanent magnet stepper motor

FIGURE 3-10   DC stepper motor (courtesy of THE SUPERIOR ELECTRIC COMPANY)

of motor is located on many manipulator drives for several reasons. It has smooth acceleration under heavy loads and does not overheat. It also has high starting torque and good running characteristics. The high starting torque is available because of the wire-wound rotor.

This type of motor does have the disadvantage of being very slow to start. This limits its application to operations that do not require quick motion on the part of the manipulator.

The induction motor has a high resistance inserted in the circuit with the rotor windings. As the speed of the motor increases, the amount of resistance is decreased. Once the motor has come up to speed, its slip rings remove the resistance box and the motor operates as an ordinary squirrel-cage motor.

Squirrel-Cage Motor

The design of the rotor resembles a squirrel cage, so the name implies a type of motor that has short-circuited conductor bars in the rotor. See Figure 3-11A. Since an AC motor is nothing more than a short-circuited transformer, the rotor or secondary is shorted and allowed to rotate. This shorted rotor, with a fan on the end of the shaft to cool it, allows it to rotate without drawing too much current and overheating. See Figure 3-11B.

The manipulator arm is usually powered by a squirrel-cage motor. There are different design characteristics to meet the particular needs of a manipulator. Most of these design characteristics concern the way the rotor is made, its speed, and the amount of current it will handle to do the job right.

The rotor for a squirrel-cage motor resembles that part of a wheel that squirrels and gerbils use to keep themselves occupied when caged up. It is made of laminated core pieces of silicon steel and has a copper or aluminum end ring and bars for conductors that run the length of the rotor.

Squirrel-cage motors are broken down into six classifications. They are listed as A through F, and each has its own characteristics that distinguish how it will operate and under what conditions. The following breakdown shows why the classifications are needed for proper choice of motor.

- Class A: Normal torque; normal starting currents; most popular type of squirrel-cage motor.
- Class B: Normal torque; low starting currents.
- Class C: High torque; low starting currents.
- Class D: Very high slip percentages; most often used for robot applications.
- Class E: Low starting torque; normal starting currents.
- Class F: Low torque; low starting currents.

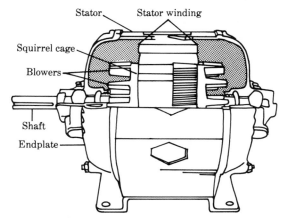

A. Cutaway view of squirrel-cage motor

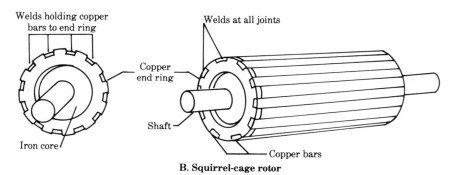

B. Squirrel-cage rotor

FIGURE 3–11   AC squirrel-cage motor

Slip

The rotor of the induction motor cannot keep up with the changing magnetic field in the pole pieces or stator. The difference between the speed of the rotor and the changing magnetic field generated by the 60 hertz alternating current is known as slip. Slip results when there is a loss of induced current between the stator poles and the rotor conductors. Load torque is what causes the motor to develop slip. The difference in speed of the rotor and the changing magnetic field is what makes the motor turn. If the rotor and the magnetic field were rotating at the same speed, the difference in the induced magnetic field and the one in the stator would inhibit the rotor from producing the torque to maintain its speed. Even with no load, the rotor tries to keep up with the changing magnetic field but cannot do so because of the friction of the bearings and the wind resistance created by the fan and the moving rotor. This is in addition to the loss in induced currents in the rotor.

Slip is indicated as a percentage. It may be from 1 percent to 100 percent. The normal slip after the motor is running is 5 percent. It will vary with the design of the rotor, the load on the motor, and what the motor was designed to do.

The motor has a tendency to increase in speed until it reaches the torque demands of the load attached to the rotor shaft. The torque will increase to meet the demands of the load. Once the torque limits have been reached, the motor has been overloaded and it stalls. This causes it to draw excess current and smoke and get very hot.

☐ END EFFECTORS

In order for a robot to do work, it must be able to pick up and place objects as well as weld, paint, or move objects. The human hand is a very complicated device that does all kinds of things that are difficult to engineer into a single mechanical unit. Most industrial robots have but one arm, and at the end of this arm (manipulator) there is some type of tooling to make it possible for the arm to pick up or move objects. This is called an end effector or end-of-arm tooling. Both terms are found in the literature, but end effector seems to be gaining acceptance as the term to be used.

The manipulator is used to move the end effector to where it is supposed to do its job. Once the end effector is at the point where it is supposed to do its work, there must be some way for it to grasp or pick up the material. And, in the case of the vacuum-operated end effector, there must be some way of turning off the vacuum to release the part once the robot has reached the programmed place. Many different designs of end effectors are available from manufacturers. They operate by using a vacuum, a magnet, or a mechanical gripper.

End effectors are divided into two groups: grippers (similar to the human hand) and end-of-arm tooling.

Grippers

The gripper performs no operation on the part it handles. It is used to pick up the part and place it somewhere else. See Figure 3–12. It may be used to place a part in a hot furnace and remove it once it is properly heated. Then it can move the piece to a bath that will treat the metal properly for a given period of time. Once the piece has been properly treated, it can be placed on a conveyor belt or on a pallet for further work at another station. The gripper is like the hand; it takes an object and holds it. The gripper may hold the object or move it while an operation is being performed on it. In either case, the gripper is designed by the manufacturer or in some cases the owner of the robot to perform a specific function.

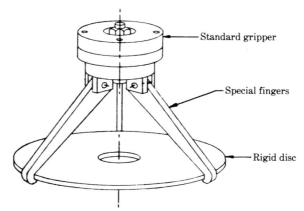

A. Three-finger rigid disc gripper

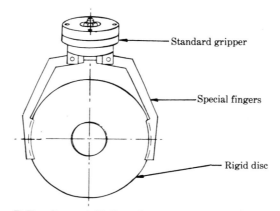

B. Two-finger rigid disc gripper

FIGURE 3–12 Two- and three-finger grippers (courtesy of MACK CORPORATION)

Vacuum Grippers

The finger-grasping type of gripper is not the only one available. Vacuum cups also are available as standard items. See Figure 3–13. They are chosen to do a job of lifting by the weight of the object to be moved. Vacuum cups are attached to the end of the manipulator, and a vacuum hose is attached to the cups. A good vacuum system will be able to cause the suction to hold the flat part until it can be moved and deposited in its desired location. The vacuum must be turned off once the part has reached its programmed place. The controller then is used to determine when the suction is applied to pick up the object and when it is to be released to allow the part to rest in its programmed space.

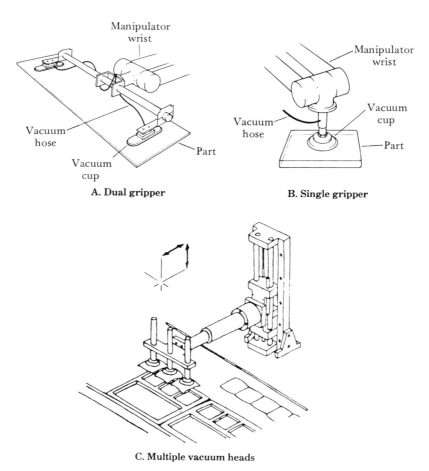

FIGURE 3-13   Vacuum grippers (parts A and B from Malcolm *Robotics;* part C courtesy of MACK CORPORATION)

Magnetic Grippers

An electromagnet may be used as an end effector. The electromagnet is designed so it can fit the object to be picked up. In this instance, the object must have a flat surface and it must be ferromagnetic. That means it has to be easily picked up by a magnet. See Figure 3-14.

End-of-Arm Tooling

If a robot is to be used for welding or spray painting, another type of end effector can be used. A cutting torch can be fitted on the end of the manipulator or a spray hose can be connected to a nozzle to allow painting to a precise degree. The end effector mounting flange holds the

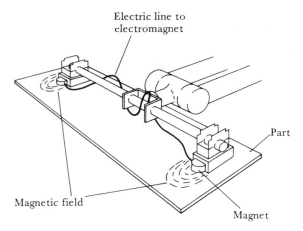

Electric line to
electromagnet

Part

Magnetic field

Magnet

FIGURE 3–14 Dual magnetic grippers (from Malcolm *Robotics*)

spray nozzle or the welding torch or cutting torch so it can be manipulated easily and precisely. The end of the manipulator usually has holes that accommodate the attachment of different types of end-of-arm tooling. See Figure 3-15. The ability of the manipulator to lift certain loads is critical to what the robot will be able to do. This payload or lifting capacity is adjusted in relation to the weight of the end-of-arm tooling. The tooling is also part of the weight that has to be lifted by the manipulator. Payload is usually given in kilograms. So, if you have an end-of-arm tooling that weighs 4 kilograms and the limit of the manipulator is 10 kilograms, it means the load that is picked up cannot weigh more than 6 kilograms.

The end-of-arm tooling is sometimes connected to the manipulator by a safety joint. The safety joint has to be designed to protect the tool in case the robot crashes.

The path for the tool is determined by the controller. The controller is very important when it comes to using the robot for welding. The path of travel has to be precise. This type of welding is done in large manufacturing plants, such as in the manufacture of automobiles and appliances. Spot welding, arc welding, and gas welding are done by robots in large numbers. These welding tools are mounted on the end of the manipulator. Refer to Figure 3-15 again.

☐ POSITIONING

One of the most important and demanding aspects of designing robots lies in the ability of the robot in *positioning* the manipulator. The robot must keep the manipulator there as long as it takes to perform the task called for by the controller.

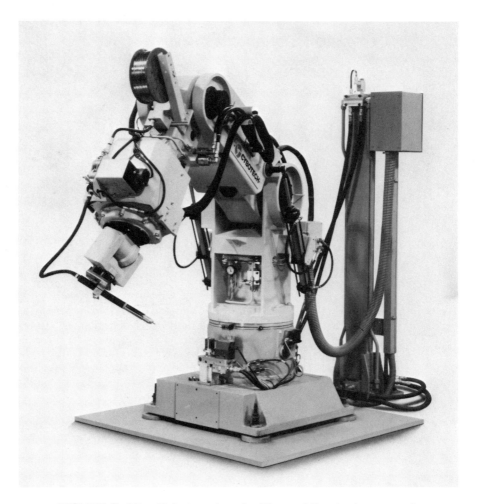

FIGURE 3-15   Robot equipped with a welding tool as an end effector (courtesy of CYBOTECH® INDUSTRIAL ROBOTS)

Robot controllers are classified as either low technology, medium technology, or high technology. These classifications allow us to indicate what level of operation the robot is able to accomplish. See Figure 3-16. (Note that not everyone in the robotics field agrees with this low, medium, and high technology classification. Therefore, the terms are used here to provide continuity of thought for this book.)

Low technology controllers:

1. are not easily reprogrammed and rely on mechanical stops to control their movements.

ELECTRO-MECHANICAL TIMERS

PROGRAMMABLE CONTROLLERS

SMALL COMPUTERS

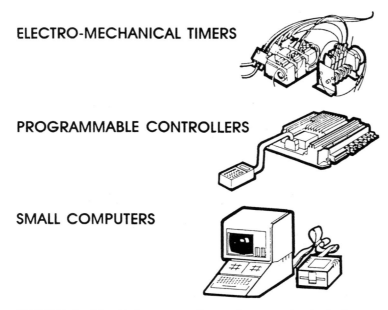

FIGURE 3-16   Robot controllers – electromechanical timers for low technology robots; programmable controllers for medium technology robots; small as well as large computers for high technology robots (courtesy of MACK CORPORATION)

2. take a long time to reprogram.
3. do not have an internal memory for storage of information.
4. do not have a microprocessor to give command signals for control of axis movement.

Medium technology controllers:

1. are used with a two- or four-axis manipulator.
2. have a microprocessor and a memory.
3. have a limited capacity for input/output signals for control of peripherals.
4. are slow to react to commands and can support movement on one axis at a time.
5. can be reprogrammed since they have a memory.
6. are usually limited in the amount of memory available (usually enough for two programs without reprogramming).

High technology controllers:

1. have a large memory.
2. have a microprocessor and a co-microprocessor.
3. have servo control for the manipulator.

4. can communicate with the peripherals with up to 64 input/output signals.
5. can be reprogrammed quickly.
6. can manipulate up to ten axes at a time.
7. have smooth operation of the manipulator arm (available in all, not just high technology types).
8. can store memory on floppy disks, magnetic tapes, and bubble memory cassettes.
9. can work with computer-aided design/computer-aided manufacturing systems.
10. can interface with sensing devices.

Controlling the action of the manipulator becomes very important in the operation of any robot. That is where the controller chosen makes the difference in the possible use of the robot. The velocity of the manipulator axes as it approaches the programmed point in the work cell is very important. This is accomplished by speed controls on the motors that operate the manipulator and its end effector. The motor is stopped at the programmed point by the use of dynamic braking or plugging.

Dynamic braking takes place when a resistor is switched across the armature of the motor as quickly as the electricity is turned off. See Figure 3–17. Current is generated by the armature in what is called a generator action. This means the armature acts as a generator once it continues to rotate and a magnetic field is still present for a short period. By putting a resistor across the armature, this current is directed in the opposite direction than that which caused it, causing the armature to be halted as the unlike polarities have a tendency to be attracted.

Plugging takes place when a motor has its electrical power reversed in polarity. See Figure 3–18. This reversal of polarity has a tendency to stop the motor immediately. It can be harmful to a motor if

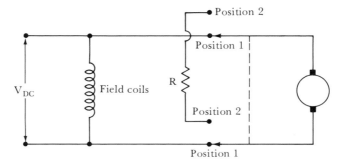

FIGURE 3–17 Dynamic braking circuit (Moving switch from position 1 to position 2 causes a reversal of the DC polarity and the motor stops quickly.) (from Malcolm *Robotics*)

done too often. It has to be done quickly and then the reverse polarity
has to be disconnected once the armature has stopped turning. Usually
the switching action is such that the polarity is returned to its original
direction after the quick application of a reverse polarity.

A stepper motor may be used with medium technology manipulators. The stepper motor has pulses sent to its windings. These pulses,
according to the design of the motor, will cause it to move a certain number of degrees. Instead of making a full 360° turn, it may rotate only 2°
with each pulse. By applying the right number of pulses, the controller
is able to make the motor step to the correct position.

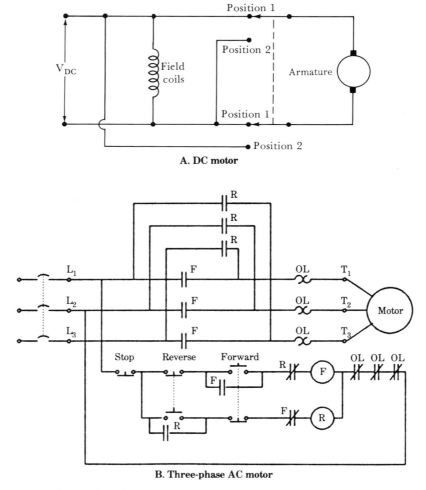

FIGURE 3-18   Plugging circuits (Switching action reverses
the polarity. Three-phase AC motors can also be plugged.) (from
Malcolm *Robotics*)

High technology manipulators use DC brushless motors. In some cases, they are driven by AC motors. The main advantage of the brushless motor is the elimination of sparks. Of course, the three-phase AC motor creates no sparks either. However, split-phase or single-phase motors do create sparks when their start switch opens after the motor comes up to speed.

The manipulator receives a signal (command) to move. It is both a positional command and a velocity command. A tachometer is used to generate a signal that is fed back to the controller to indicate the speed at which the manipulator is moving. The tachometer is nothing more than a simple DC generator. It resembles the permanent magnet electric motor, except that it is driven by the turning of the manipulator and thus generates a signal as its armature rotates in the permanent magnet field. They are similar to the permanent magnet motors used to drive toys, except they are made with a little more precision.

☐ REPEATABILITY AND ACCURACY

One of the most important characteristics of any robot is its ability to do the same thing over and over again accurately. The robot is programmed to do a specific task. It is led to the place where this task is to be performed. This route to and from the task is programmed into its memory. The location of the points is stored and are recalled each time it is needed to repeat the job. After the programming has been completed and the "run" command has been given, the robot may not return to the exact spot. It may miss the spot by 0.030 inch. If this is the greatest error by which the robot misses its point, then its *accuracy* is said to be ± 0.030 inch.

Medium technology robots do not have the accuracy and repeatability of low technology robots. This is due to the increased number of axes on medium technology robots. They have several axes that must converge on a point. Errors created by the many axes are cumulative and add up to less accuracy. Low technology robots move but one axis at a time to reach their programmed position. Low technology robots rely on the hard stop for accuracy. Accuracy of medium technology robots ranges from 0.2 millimeter to 1.3 millimeters.

Accuracy is not the only important characteristic of a robot. The robot's repeatability is also important in doing a job correctly. *Repeatability* is a measure of how closely a robot follows its programmed points and returns every time the program is executed. The robot may miss its programmed points by 0.030 inch the first time the program is executed. During the next execution of the program, if the robot misses the point it reached during the previous cycle by 0.010 inch, or a total of 0.040 inch from the original programmed point, its accuracy is 0.030 inch and its repeatability is 0.010 inch.

Repeatability can change with use and time. The mechanical components have a tendency to wear that increases the inaccuracies of the whole system. Most inaccuracies are easily corrected. In some applications, good repeatability is more desired than accuracy.

☐ DRIVES

Gears, belts, and chains make up the drive systems of robots. Each has its particular application. The main reason for these drives is to transfer energy. In order to move the manipulator and the end-of-arm tooling, some means of transmitting energy to both must be devised. That is where the chains, belts, and gears become useful. The big task is to transfer the energy from an actuator to the manipulator. The actuator is the motor or drive energy source.

Gears

The most popular method for transferring energy from an actuator to an end effector is gearing. That means if the manipulator actuator is located at the center of the manipulator, some method must be employed to move the energy generated at the actuator to the end effector. Gearing is but one method, but the one most often used.

Gears are made in many sizes and with different numbers of teeth. They can be designed to give mechanical advantages and to reduce speed. The speed of an electric motor is most often too high for the direct application to robot motions. That means it has to be geared down to a speed needed for the operation. Gears can be used to increase or decrease the effective speed of an electric motor.

Gears can be purchased in many different configurations. They can operate at right angles to one another, or they can operate one within another. They have the ability, when properly arranged, to change rotary motion to linear motion by driving a straight bar with gear teeth (called a rack). End effectors use different gear mechanisms and so do the drive portions of robots. There are many different grippers and finger arrangements. They too use a variety of gears and levers to convert the power source into the desired motion and power needed to perform the job.

Gear Trains

When several gears are placed together, they form a train. See Figure 3–19. This is done in order to change the direction of rotary motion or to increase or decrease the speed of rotation of the final gear shaft. There are two types of gear trains: ordinary gear trains and planetary gear trains. The ordinary gear train is broken down into two other classifications: simple gear train and compound gear train.

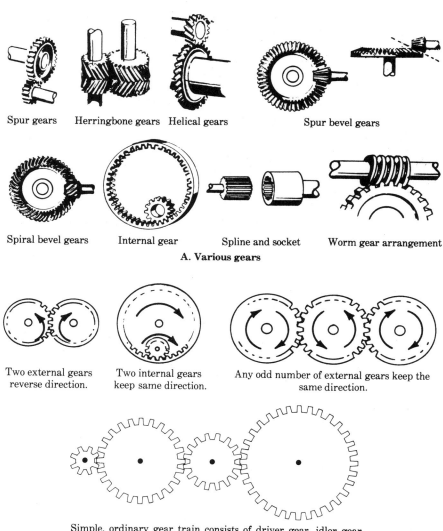

Spur gears   Herringbone gears   Helical gears   Spur bevel gears

Spiral bevel gears   Internal gear   Spline and socket   Worm gear arrangement

A. Various gears

Two external gears reverse direction.   Two internal gears keep same direction.   Any odd number of external gears keep the same direction.

Simple, ordinary gear train consists of driver gear, idler gear, idler gear, and driven gear.

B. Gear trains

FIGURE 3–19   Gear arrangements and gear trains

Worm Gears

Worm gears are often used for driving the base of a manipulator in robots. See Figure 3–20. There are some complex arrangements for making sure the right speed and direction are obtained. The worm is the *screw* with either a single thread or multiple threads. The form of the axial cross section is the same as that of a rack. The teeth of the worm

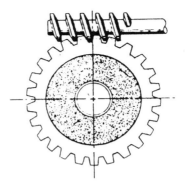

FIGURE 3-20   Worm gear (The wheel part is round, and the
worm part is the threaded rod-shaped part.)

wheel have a special form that is required to provide proper conditions
for meshing with the worm.

    Worm gears are used to convert a circular or rotary motion to a
circular motion. This is important when it comes to turning the manipulator arm of a robot. An electric motor, a pneumatic motor, or a hydraulic motor may be used to drive the worm gear arrangement to power the
manipulator.

Ball Screws

When servo motors and stepping motors are used in robots, it is often
necessary to convert the rotary motion to linear motion. This was done
for years by a threaded rod with a nut. If the nut is kept from turning as
the threaded rod is rotated, the nut will move along the threaded rod.
The major problems with this system were friction and wear. When the
threaded rod was turned in the nut, the high friction produced by the
tightly fitting screw threads wasted much of the power created by the
motor. As the rod turned in the nut, the nut and threaded rod would
wear. This wear created inaccuracies that had to be dealt with before
this method could be used in robots and the more modern automated
systems.

    These problems have been overcome with the improvements in
the ball bearing screw drive. See Figure 3-21. The *ball screw* has been
around for many years. The groove in the screw is similar to the groove
in a standard screw but is cut to allow a ball bearing to roll freely in the
groove. Some automobiles now use the ball bearing screw drive method
for steering.

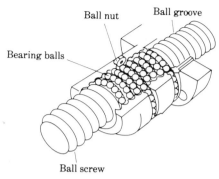

A. Cutaway view of ball screw

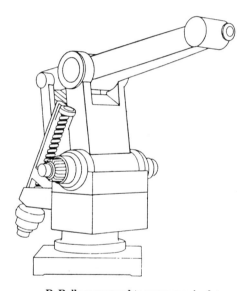

B. Ball screw used to move manipulator

FIGURE 3-21 Ball screws (part A from Malcolm *Robotics;* part B courtesy of WARNER CLUTCH AND BRAKE COMPANY)

Bevel Gears

Bevel gears are conical (cone-shaped) gears and are used to connect shafts that have intersecting axes. See Figure 3-22. Hypoid gears are similar to bevel gears in their general form, but they operate on axes that are offset. Most bevel gears can be classified as either the straight-tooth type or the curved-tooth type. Spiral bevels, Zerol bevel, and hy-

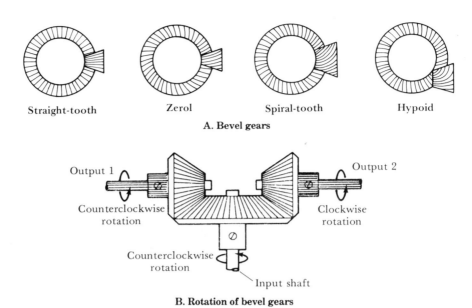

Straight-tooth   Zerol   Spiral-tooth   Hypoid

A. Bevel gears

Output 1 — Counterclockwise rotation

Output 2 — Clockwise rotation

Counterclockwise rotation — Input shaft

B. Rotation of bevel gears

FIGURE 3–22   Bevel gear arrangements and rotation (from Malcolm *Robotics*)

poid gears are all classified as curved-tooth gears. Straight bevel gears are the most commonly used type of all the bevel gears. However, most of these are not used on robots. The teeth are straight, but their sides are tapered so that they intersect the axis at a common point (called the pitch cone apex) if they were extended inwardly.

Bevel gears are used in robot manipulators when a 90° or 45° transfer of motion must take place. The cut of the tooth in bevel gears determines the ability to withstand heavy loads. Spiral-tooth gears can handle heavier loads than Zerol cut gears. The hypoid gear teeth and that of the Zerol are similar. The only difference is that the driver gear is offset from the center contact point of the gear. This gearing arrangement is sometimes referred to as the mitre gear. That is because the 45° and 90° arrangement appears the same as a mitre joint.

Adjusting Gears

Ball screw drives convert the rotary motion of a servo motor to linear motion. Many robots require both linear and rotary motion. A good example are the joints of a jointed-arm robot. They can be driven with a rotary drive.

Most servo motors cannot be directly connected to the joint of the robot to produce rotary motion. The power developed by most servo motors is not high enough to drive the arm of the robot, especially when

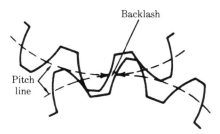

FIGURE 3-23 Backlash—the looseness between gear teeth when they mesh

the shaft of the servo motor is connected directly to the robot's joint. If the armature of the servo motor is connected to a transmission and the output of the transmission to the robot's joint, there will be sufficient power to move the robot's arm. It will also have enough power to move the load that the robot is carrying.

The servo motor turns at high speed to drive a transmission. The servo motor turns the input shaft of the transmission at high speed. Gears in the transmission reduce the speed and increase the torque. The output shaft of the transmission turns at low speed and high torque. This high torque and low speed are sufficient to drive the robot arm.

The problem arises when gear teeth mesh with one another. When they meet one-on-one, they have a tendency to wear. This wear is accelerated by the speed of rotation of the gears. As the gears wear, the output shaft is driven less and less evenly. The robot arm no longer moves smoothly. That means positioning of the arm becomes inaccurate.

Backlash is the looseness in the gears where they mesh. See Figure 3-23. That means the amount of tooth space exceeds the thickness of the engaging tooth. As gears wear, the backlash becomes greater. Backlash can be eliminated or at least reduced by bringing the gears closer together. As the backlash is removed, the friction between the gears is increased. This increased friction wastes power and generates heat and noise in the transmission. There must be enough backlash to allow for lubrication to coat the gear teeth. Machinists' handbooks list the correct backlash for gears to run efficiently and quietly.

New drives are being developed and will be encountered on some of the latest robots. Whenever there is a demand, there is an answer. New products are constantly being developed.

☐ HARMONIC DRIVES

Harmonic drives are used in robots to reduce backlash, gear wear, and friction. They have been developed to improve the quality of the trans-

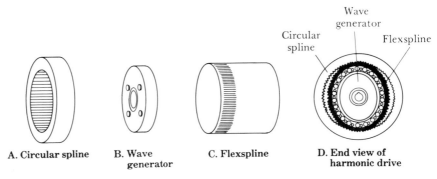

FIGURE 3-24   Harmonic drive (from Malcolm *Robotics*)

missions used in robots. The *harmonic drive* consists of three major elements: circular spline, wave generator, and flexspline. See Figure 3-24.

The wave generator is an ellipse. The flexspline is a flexible cup with teeth cut on the outside diameter. The wave generator is slid into the flexspine, distorting it into an ellipse. The circular spline is a nonflexible ring with gear teeth cut on the inside diameter. The wave generator and flexspline assembly is slid into the circular spline.

Gear reduction is generated when the harmonic drive wave generator is rotated. The ratio of the harmonic drive is controlled by the number of teeth on the circular spline and the difference in number of teeth between the circular spline and the flexspline. For example, if the circular spline has 400 teeth and the flexspline has 398 teeth, the ratio is 400:2, or 200:1. If the circular spline has 100 teeth and the flexspline has 97, the ratio is 100:3, or 33.34:1. The possibilities are limited only by the number of teeth that can be put on the circular spline and flexspline.

One of the greatest advantages of harmonic drives over standard transmissions is the total lack of backlash. This alone increases robot accuracy and improves its efficiency.

☐ BELTS

Sometimes it is not possible to use gears to transfer power from the actuator to the robot. Other methods must be used. Chains and belts are the next best possibilities. Belts are flexible and quiet running. They are also able to absorb some of the shock produced by the stop-and-go operations of a robot. Belts do, however, have some limitations, inasmuch as they are flexible and subject to wear. By far, the greatest limitation is slippage. They can be reinforced and made useful in many operations.

Three types of belts can be considered for use with robots: V-belts, synchronous belts, and flat belts.

V-Belts

V-belts get their name from their shape. The *V* fits into a pulley rather easily. See Figure 3–25. A closer examination reveals that the belt is made of rubber with reinforcement cords running throughout its length.

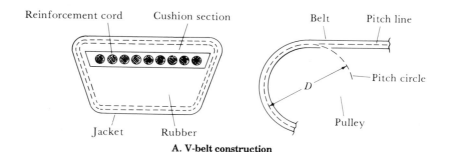

A. V-belt construction

Fractional Horsepower		Standard Multiple		Wedge	
Size	Dimensions	Size	Dimensions	Size	Dimensions
"2L"	1/4 — 5/32				
"3L"	3/8 — 7/32				
"4L"	1/2 — 5/16	"A"	1/2 — 5/16	"3V"	3/8 — 5/16
"5L"	21/32 — 3/8	"B"	21/32 — 13/32		
		"C"	7/8 — 17/32	"5V"	5/8 — 17/32
		"D"	1-1/4 — 3/4		
		"E"	1-1/2 — 29/32	"8V"	1 — 7/8

B. Various sizes and shapes of V-belts

FIGURE 3–25 Construction details of V-belts (from Malcolm *Robotics*)

This type of belt is connected to a motor pulley and a pulley on the base of the manipulator.

Synchronous Belts

Synchronous belts are also recognized by their shape. They have evenly spaced teeth where they contact the pulley. The pulley is also specifically designed to go with this type of belt. See Figure 3–26. That means the synchronous belt is more expensive than the V-belt. The teeth of the belt mesh with the grooves in the pulley to make sure there is no slippage. This type of belt is used to provide a positive grip to the robot's manipulator wrist assembly. It is used where there is a constant change of direction of rotation. Synchronous belts can also be found on some automobile engines; they are known as timing belts.

Flat Belts

Flat belts are used in the wrist assembly in small manipulators because they have a tendency to slip when large loads are placed on them. The

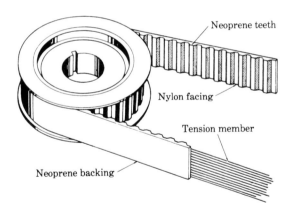

Neoprene teeth

Nylon facing

Tension member

Neoprene backing

FIGURE 3–26   Construction details of a synchronous belt

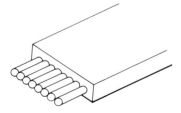

FIGURE 3–27   Flat belt reinforced with cords (from Malcolm *Robotics*)

flat belt is made of rubber and reinforced with cords along its entire length. It is inexpensive and therefore ideal for transferring power from one source to another. It is good for low and moderate speeds to deliver large torque when needed. See Figure 3–27.

☐ CHAINS

Chains are used when belts will not do, when you need a source of energy transfer that will not slip and can handle large loads. They come in handy for long-distance (longer than gears and shorter than belts) transfer of energy. Chains do not stretch or slip like belts. Roller chains, not bead chains, are usually used for robots.

Roller Chain

Roller chains provide high torque transmission and good precision. The roller is the same type of chain used on bicycles. See Figure 3–28. This type is usually utilized in transferring energy from the actuator and drive mechanism to the manipulator.

Bead Chain

Bead chains are used to close drapes and turn basement lights on and off. They are suited for low torque applications but are not suitable for robots. Bead chains can be used to drive devices with low torque requirements. See Figure 3–29. The chains break easily. The beads are made of metal or plastic. (Note from the figure how they fit into the sprocket dimples.)

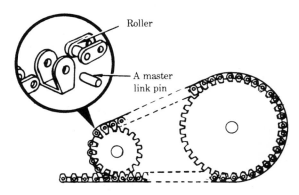

FIGURE 3–28   Roller chain

FIGURE 3-29   Bead chain (from Malcolm *Robotics*)

☐ SUMMARY

Robots need some type of system to cause them to function. Hydraulic, pneumatic, and electric drive systems are all utilized to drive robots.

Hydraulic systems are used for heavy loads. Pneumatic systems are used for medium and low load weights. The electric drive is used for low load weights.

Hydraulic systems use pumps to create the flow needed to do the work at the end of the arm. There are several types of hydrostatic pumps used to make the system operate properly.

Pneumatic systems use air to do work. There are pneumatic motors used on the end of the manipulator to grip or handle the load being processed. The pneumatic system does not need a return system for the air. It exhausts directly into the atmosphere.

Electric drive systems are powered by electric motors. There are many types of electric motors, but the DC types are preferred for precision motion and movements. Permanent magnet, stepper, brushless DC, and the Hall-effect DC motors are used for various functions in robots. AC motors are used for heavy loads and where precision of movement is not necessary. They are classified as induction and squirrel-cage types. Each type has its particular applications.

Squirrel-cage motors are further broken down into six classifications according to their starting currents and torque.

End effectors may also be called end-of-arm tooling. (The terminology has not been agreed upon at this time.) The manipulator is used to move the end effector that is mounted on the end of it. Grippers are used to pick up and hold objects being machined or boxed or picked up and placed or palletized. There are vacuum-operated grippers and magnetic grippers as well as a variety of mechanical devices used to grip or hold materials. Many of these grippers are made in the plant where the robot is working.

Positioning is very important in the proper utilization of a robot. The robot must be able to place an object in the same location over and over again without being too far off the spot.

Controllers, and consequently robots, are classified as low technology, medium technology, and high technology. The ability of the controller to handle programs for the robot makes the difference in its classification.

Dynamic braking and plugging are both used to promptly stop the movement of the manipulator. They both have advantages and disadvantages and can be used in various locations depending on the application.

Repeatability and accuracy are important parts of the robot system. Repeatability is the ability of the robot to place an object in the same place repeatedly. Accuracy is the degree of accuracy with which it can handle the repeat function.

Gears, chains, and belts are the types of drives used in robots. Each has its own applications. Gears are accurate and noisy. Chains have their limitations, whereas belts can be used under circumstances where power is transmitted only short distances. Harmonic drives have been developed to reduce backlash and improve the efficiency of the robot operation. They have also eliminated a lot of noise inherent in standard transmissions. The ball screw is a useful adaptation of principles to eliminate backlash or gear looseness.

☐ KEY TERMS

accuracy   the degree to which a robot can place an object in a given spot repeatedly

backlash   the looseness in the gears where they mesh

ball screw   a method of using ball bearings to substitute for screw threads (The ball screw changes rotary motion to linear motion.)

brushless motor   a DC motor that operates without using brushes (An electronic circuit controls its field excitation.)

harmonic drive   a type of drive that uses a flexspline, circular spline, and wave generator to accurately position a manipulator with no backlash and little noise

hydraulics   the use of pressure on a fluid to drive an end effector or a manipulator

plugging   a method of stopping an electric motor by reversing the polarity of its power source

positioning   the ability of a robot to place a particular object in a desired location

pumps   devices, usually electrically driven, used to increase the pressure on a hydraulic fluid

repeatability   the ability of a robot to place an object in the same spot repeatedly

roller chains   the same type of chain used in bicycles; used to drive manipulators and end effectors

stepper motors   DC motors whose shaft rotates a specific amount of mechanical degrees each time its input is pulsed (The shaft is stepped from one position to the next by the proper combination of fields in the motor.)

synchronous belts   belts that have teeth that fit a pulley with grooves so they do not slip

V-belts   belts used to drive a manipulator; shaped to fit a *V* pulley

worm gears   a method of changing linear motion to rotary motion or vice versa

☐ QUESTIONS

1. List three types of robot drive systems.
2. Why are filters needed in hydraulic systems?
3. How does a relief valve function?
4. What are the two classifications of hydrostatic-type pumps?
5. What does pneumatic mean?
6. List the parts of a pneumatic drive system.
7. Describe a PM motor.
8. What is a stepper motor used for?
9. How do DC motors work without brushes?
10. What are the two types of AC motors most often used on robots?
11. Where is a squirrel-cage motor used?
12. List the torque characteristics of the six classifications of squirrel-cage motors.
13. What is slip? How is it used to advantage?
14. How do vacuum grippers work? What are the limitations of vacuum grippers?
15. How do magnetic grippers work? What are the limitations of magnetic grippers?
16. What is another name for end-of-arm tooling?

17. What is positioning?
18. What is repeatability?
19. What is meant by robot accuracy?
20. What is the advantage of a roller chain drive?

ROBOT SENSORS

R obots are made with as many human qualities as possible. They must have the five senses humans possess – sight, hearing, taste, touch, and smell. Some of these senses are used to make robots do special tasks. In most instances, the robot does not need to smell. Of all the senses, sight is the most difficult to perfect. Vision for robots is discussed in another chapter.

□ SENSORS AND SENSING

In order to pick up things, the robot must be able to sense when it has the object gripped tightly enough to hold. On the other hand, the robot does not want to crush the object while trying to pick it up. Therefore, there must be some type of sensing device in the gripper or attached to it that regulates the amount of pressure applied to the object being retrieved.

Transducers are used to convert nonelectrical signals into electrical energy. A transducer, then, can serve as a sensor. It converts the pressure applied to it into a signal that is fed to the robot controller. The controller then takes this signal strength and uses it to increase or decrease the pressure being applied by the gripper or end effector. See Figure 4–1.

Limit switches are designed to be turned on or off by an object hitting a lever or roller that operates the switch. See Figure 4–2. Robots that do repetitive jobs with little demand for complex operations use limit switches to tell them when to reverse or stop. These switches are either on or off and are very simple in their operation.

If a robot is to make decisions and assemble products, it needs force sensors, vision sensors, and tactile (feel or touch) sensors. As the types of sensors on the robot increase, its ability to do complicated processes increases.

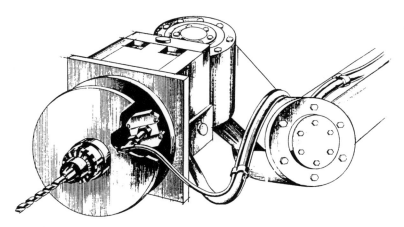

FIGURE 4–1   Piezoresistive transducer on end effector used to monitor pneumatic pressure for cost-effective control of automated drilling (courtesy of MICROSWITCH, a Honeywell Division)

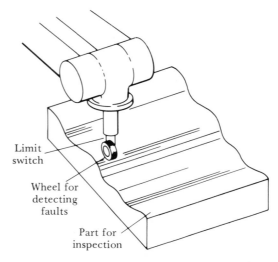

Limit
switch

Wheel for
detecting
faults

Part for
inspection

FIGURE 4-2   Limit switch (from Malcolm *Robotics*)

☐ CLASSES OF SENSORS

Sensors may be classified as either contact or noncontact. They may be further classified as internal or external and as passive or active.

Contact Sensors

A limit switch is a *contact sensor.* This limit switch permits the robot to sense whether an object is present or missing. See Figure 4-3. If the object makes contact with the limit switch, then the robot knows the object is near enough to begin its operation. If the switch is not closed, it means the object is missing and the robot has to react accordingly. That usually generates what is referred to as an alarm condition.

Force, pressure, temperature, and tactile sensors all respond to contact. They all send their signals to the controller for processing.

Noncontact Sensors

Pressure changes, temperature changes, and electromagnetic changes can all be sensed by noncontact methods. They usually react to a change in a magnetic field or light pattern. If an electromagnetic field is disturbed, it is sensed and fed to the controller. The same is true of the disturbance of a light beam. Changes in the light beam – its intensity or whether or not it is present – are sent to a controller for processing and then sent back to the robot to react according to the disturbance. See Figure 4-4.

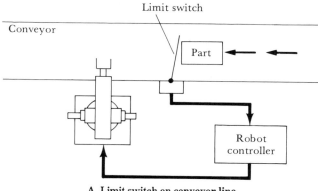

A. Limit switch on conveyor line

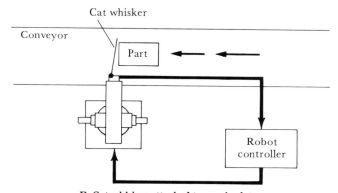

B. Cat whisker attached to manipulator

FIGURE 4–3   Contact sensor (from Malcolm *Robotics*)

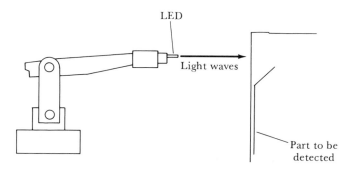

FIGURE 4–4   Noncontact sensor (from Malcolm *Robotics*)

A number of *light-emitting diode (LED) sensors* are used in robotics. The LED produces a low-level light beam that is picked up by another device. If the light beam is broken, the part to be picked up by the robot is present. If the light beam is not broken, then the robot goes into a different mode for which it has been previously programmed. See Figure 4–5.

Another noncontact sensor is the television camera mounted on the end of the manipulator. It can see the presence of parts and compare them to what is in the memory of the computer and then pick up the right part and move it to a preprogrammed location. See Figure 4–6.

Self-Protection

The robot must be protected from any damage it may do to itself as well as the injury it may cause to the human worker within its work envelope. Properly mounted sensors make sure the arm of the robot does not move where it is not programmed to be. If this does occur, the robot comes to an emergency stop and will remain stopped until the operator starts the program again. Figure 4–7 shows a mat that contains switches that prevent the robot from starting while a person is inside the work envelope where the mat is located.

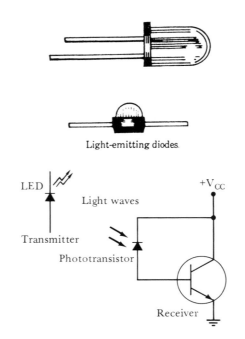

Light-emitting diodes.

FIGURE 4–5   LED sensor (from Malcolm *Robotics*)

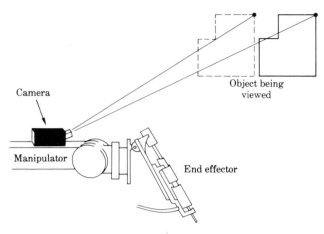

FIGURE 4-6 TV camera on a manipulator (from Malcolm *Robotics*)

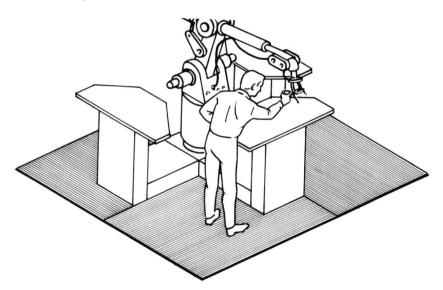

FIGURE 4-7 Flexible ½ " thick mats with tapeswitches (courtesy of TAPESWITCH CORPORATION OF AMERICA)

As previously mentioned, the work envelope is that area where the manipulator moves the end effector. It varies according to the robot and its design characteristics. See Figure 4-8. Robots are usually located in a cage to prevent humans from entering the work envelope. The quick movements of the robot arm may cause serious injury to anyone it hits. Maintenance persons have to be very careful whenever they are in the work area of the robot.

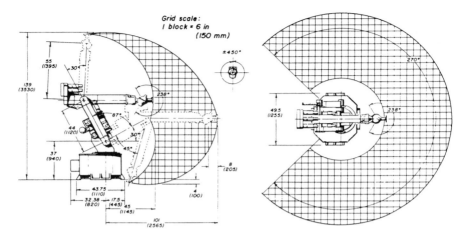

FIGURE 4-8   Work envelope (courtesy of CINCINNATI MILA-CRON)

Worker and robot safety are of primary importance in any installation where robots are used. The robot can be programmed, but it takes some time to train the human worker to be careful when within the robot's domain.

Collision Avoidance

The biggest challenge a programmer has in programming a robot is *collision avoidance* with the object to be serviced. If the gripper hits the part to be picked up and knocks it out of place, then the gripper will not be able to locate it and the whole operation will have to be stopped until the right program is fed into the controller. Irregular shapes are difficult to program, and it takes time to develop a proper program for them. Proximity sensors can be used to sense the presence or absence of the object. See Figure 4-9.

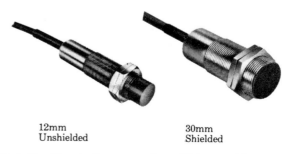

12mm
Unshielded

30mm
Shielded

FIGURE 4-9   Proximity sensor for collision avoidance

□ PROXIMITY SENSORS

Proximity sensors give the robot the senses of touch and sight. They take various forms and work on different principles.

Using electronic commands, a proximity sensor can send out signals indicating the proximity of the part being processed. The inductive proximity sensor is based on an *LC* oscillator circuit. (*LC* stands for inductive–capacitive.) The oscillator sets up a frequency that is turned off when a metallic object comes near it. The metallic object changes the inductance of the circuit and the frequency. The amplifier amplifies the signal and then processes it to turn a switch on or off according to what is the preferred operation at this point.

The *RC* circuit is another type of electronic proximity sensor. (*RC* stands for resistive–capacitive.) This oscillating circuit has its frequency changed by the closeness (proximity) of the object being sensed. The signal is processed by an amplifier, and a switch is turned on or off according to the preferred operation at this point.

An advantage of the resistive–capacitive sensor is its ability to detect metallic or nonmetallic objects.

Pulsed infrared photoelectric controls are used in industrial robotics to sense the presence of any type of object. These devices can detect the product on the in-feed or out-feed conveyor to and from the protected robot work station.

Eddy current proximity detectors use magnetism to function. They induce a magnetic field in any object nearby. A small coil picks up any change in the magnetic field about it and sends it on to the controller for processing. The reed switch is another magnetic electrical proximity switch that responds to a controlled magnetic field. It makes or breaks contact when exposed to either a permanent magnetic field or an electromagnetic field. The contacts of the switch are inside a hermetically sealed glass tube. The two flat metal strips or reeds are housed in a hollow glass tube filled with an inert gas. When the magnetic field is brought near, the reeds are pulled together and make the contact needed for the work to be done. In some instances, the magnet may cause the switch to open instead of close, depending on the type of switch needed for the job. Figure 4–10 shows a reed switch.

FIGURE 4–10   Reed switch

☐ RANGE SENSORS

If precise distance is needed, it is best to use a range sensor. *Range sensors* are used to locate objects near a work station to control a manipulator. One range sensing system is the laser interfero-metric gage. It is very expensive and is sensitive to humidity, temperature, and vibration. A television camera is another range sensing system.

☐ TACTILE (TOUCH) SENSORS

Tactile sensors rely on touch to detect the presence of an object. The two types of tactile sensors are touch sensors and stress sensors. Touch sensors respond only to touch. Stress sensors produce a signal that varies with the magnitude of the contact made between the object and the sensor.

The simplest touch sensor is the microswitch, which is turned on or off by the presence of an object. See Figure 4–11. The strain gage is a good example of a stress sensor. Figure 4–1 shows a good use for a sensor in the end effector.

Strain Gage

A *strain gage* is used to sense mechanical movement. This device can be used with grippers to determine the amount of force being applied to an object with the grasp of the mechanical fingers. See Figure 4–12.

FIGURE 4–11   Microswitches used as touch sensors

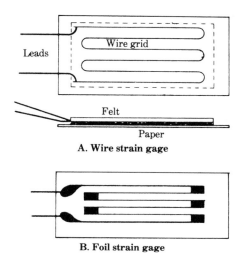

A. Wire strain gage

B. Foil strain gage

FIGURE 4–12 Strain gages

Older strain gages were made of fine wire (about 0.001 inch in diameter). The wire was attached to an insulating strip of material. As the unit was stressed, the wire would stretch. As the wire got longer, its cross-sectional area was reduced and its length was increased. This changed the resistance of the wire, thereby changing the current through the circuit.

Newer strain gages are made of semiconductor materials and provide greater sensitivity. They change resistance approximately fifty times faster than metallic gages and have a higher output.

Pulsed Infrared Photoelectric Control

These controls are used to sense the presence of any type of object. They detect a product on the in-feed or out-feed conveyor. They can also be used to detect the presence of an object to and from the protected robot work station. See Figure 4–13. These devices will operate in any environment. They disregard ambient light, atmospheric contamination, and thin film accumulations of oil, dust, water, and other airborne deposits. Low voltage is used from the sensor to the controller.

About 35 percent of all robots are used for pick-and-place operations, so the robot's gripper is very important. The gripper must be able to pick up an object and hold it until it is commanded to release it. There also may be requirements as to when and under what conditions the object should be released, which is where tactile sensing comes into play. The robot must be able to respond to various pressures and act accordingly.

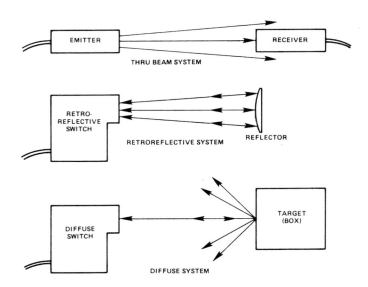

Thru-beam units have the source in one enclosure and the receiver in another, pointing at each other. These types offer the longest ranges or highest gains and are generally used in extremely dirty environments as they can tolerate the most attenuation and still function. However, since alignment is critical for peak performance and two devices must be purchased and installed, this system is more expensive than retroreflective or diffuse types.

Retroreflective units have the source and detector in one enclosure and operate by bouncing the light beam off a reflector or reflective tape. These are less expensive and easier to install than a thru-beam unit but since the beam must travel twice as far, ranges and gains are less. They are not as good as thru-beams for dirty areas and care must be taken that the target is not shiny or it will look like a reflector to the switch and be ignored.

The construction of **diffuse** types is similar to retroreflective units but they detect the small amount of light that is reflected diffusely from non-shiny targets such as cardboard boxes. Sensing ranges and gains are lower than the thru-beam or retroreflective types and dependent on the reflectivity of the target, with published values based on a target of quality white paper. While easiest to align, these units can detect objects beyond the line of travel of the intended target such as a wall or passer-by and this must be considered.

FIGURE 4-13   Infrared photoelectric control (courtesy of SQUARE D COMPANY)

Figure 4-14 illustrates a simple sense gripper. The Microbot Mini-Mover 5 has a drive motor, tension switch, idler pulleys, tension spring, torsion spring, housing, and links. Force is applied to the end of the gripper. The gripping action is cable driven. A tension spring supplies passive compliance. (Compliance is the ability to trade off on position and force.) A small stepping motor powers the gripper so passive compliance is important. The tension spring provides the passive compliance. The tension switch signals the computer that the gripper is firmly closed. In other words, as the fingers encounter force, the cable tightens, straightens, and pushes the switch till it closes.

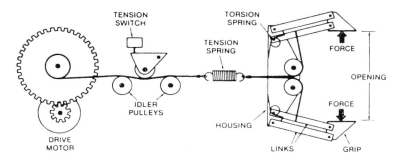

FIGURE 4–14   Gripper on Microbot's Mini-Mover 5 (courtesy of MICROBOT®)

The gripper may require more dexterity if odd-shaped objects are to be gripped. This is provided by pressure-sensitive materials mounted on the gripping surfaces. The surfaces are then divided into a grid. The computer can sense the balance of force over the contacting surfaces. A good example of such a touch sensor is the Hillis touch sensor. See Figure 4–15. It is useful in picking up nuts, bolts, flat washers, and lock washers that are identified by touch. The sensor is made of anisotropically conductive silicone (ACS). ACS is electrically conductive only along one axis in the plane of the sheet. The rubber sheet (ACS) and a flexible printed circuit board are separated by a nylon mesh. As increasing pressure is applied to an area of the ACS sheet, more of the sheet contacts the flexible printed circuit board, and the resistance of that region goes down. The flexible printed circuit board is etched with parallel conducting lines at 90° to the conductive direction of the ACS sheet. A second circuit board is used as a contact board for driving columns along the conductive lines of the ACS sheet. The result of this arrange-

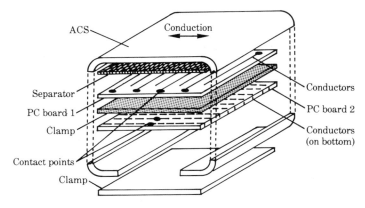

FIGURE 4–15   Hillis touch sensor

ment is a sensing grid with a resolution of approximately 1 square milli-meter, matching the sensing capability of the human fingertip.

One advantage of these touch sensors is that the computer can respond to them better, in some instances, than it can to a TV camera since the information does not have background to contend with. A dis-advantage is that these touch sensors can be damaged easily. The piezo-electric touch sensor is another tactile sensor that is becoming more common. (See Figure 4–1.)

□ TEMPERATURE SENSING

Temperature changes are easily sensed with the devices presently avail-able. They either increase resistance with an increase in temperature or decrease resistance with an increase in temperature. They may also give off a small (usually in millivolts) change in voltage as a result of the ap-plication of heat. This very low voltage can be amplified to be used in the computer and fed to the controller.

Thermistors respond to changes in temperature by increasing or decreasing resistance. See Figure 4–16. If the temperature is increased, the thermistor decreases its resistance, thereby increasing the current in the circuit. If the thermistor temperature is decreased, the resistance of the thermistor increases, thereby decreasing the current in the cir-cuit. This increase and decrease is usually not linear, so the information fed to the computer must be equated with the response curve of that particular thermistor.

Thermocouples have been used for years as temperature sensors. They are made of two dissimilar metals, such as iron-constantan, cop-per-constantan, or platinum-rhodium. See Figure 4–17. The ends of the metals are fused together. When heated, they give off a few millivolts that can be amplified and fed into a computer for processing. As the temperature increases, the output of the thermocouple increases.

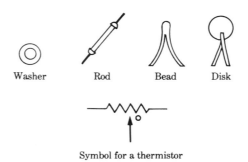

Washer   Rod   Bead   Disk

Symbol for a thermistor

FIGURE 4–16   Thermistors

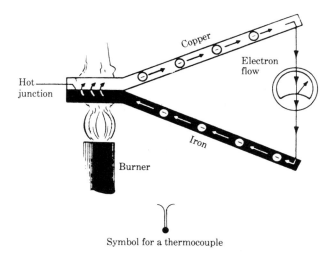

Symbol for a thermocouple

FIGURE 4-17   Thermocouples

☐ DISPLACEMENT SENSING

It is important to know the exact location of the gripper or manipulator so this information can be fed into the computer or controller for action. A number of methods are used to tell the computer or controller where a particular unit is at a specific time. One such method is using a resistive sensor, which has a fixed resistance – usually a wire-wound resistance – with a slider contact. As force is applied to the slider arm, it changes the circuit resistance. This in turn changes the amount of current flowing in the circuit. This change can be equated with location, and the computer can act on it to determine the exact placement of the arm or gripper. See Figure 4-18A.

Capacitive sensing is another method of detecting location. A capacitor is made up of two plates. If one is held stationary and the other is allowed to move, then it is possible to use this varying capacitance to change the frequency of an oscillator. The change in oscillator frequency can be detected and fed into the computer for processing in regard to the location of the manipulator or gripper. See Figure 4-18B.

Another way to detect the change in location of any device is to place an inductive sensor on it. The moving force of the manipulator or gripper will cause the movable iron core to change its location. As this metal core changes, it changes the inductive reactance of the coil. This inductive reactance, when divided into the voltage applied across the coil, determines the current in the circuit. As the current changes due to the inductive reactance change, the movement that caused the change can be detected and processed by the computer and fed to the controller for its action. See Figure 4-18C.

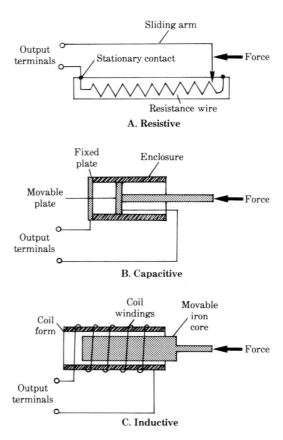

FIGURE 4-18   Displacement sensors

Other types of location or displacement sensors are LVDTs, RVDTs, encoders, pots, and resolvers, to name just a few.

□ SPEED SENSING

To detect how fast a motor or shaft is turning, some method of reliable sensing must be used. Two methods employed today are the tachometer and the photocell.

The tachometer is nothing more than a permanent magnet motor being driven as a generator. Its output is attached to a meter circuit, and the speed is read out in terms of the current generated. The motor is attached directly to the device being monitored, or it can be driven by a belt arrangement using a synchronous belt and pulleys. Speed sensing is also used in the control of a robot. See Figure 4-19.

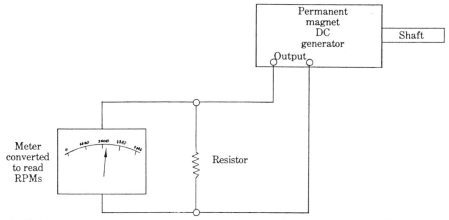

A. The tachometer unit contains generator, resistor, and meter movement calibrated to read RPMs.

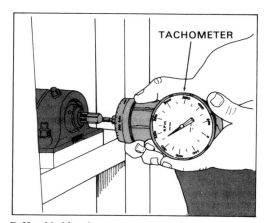

B. Hand-held tachometer can be used to check accuracy of permanently installed unit.

FIGURE 4–19 Tachometer

The photocell monitor for speed works with a light beam that is reflected by a spot on the rotating shaft. The reflective spot on the shaft causes the light to be disrupted. The sensor then reacts to this interruption and produces an electrical output in relation to the number of interruptions per second. See Figure 4–20. A strobe light is another light source for sensing speed.

A. Small magnets can be placed on pulley and
   counted as they cause a pulse to be
   generated in the stationary coil. Or, a
   transistorized "Hall effect" can be used to
   detect the speed of the motor.

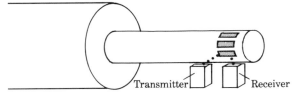

B. Beam is transmitted and hits shaded spots on the rotating shaft; reflected beam is picked up by receiver
   and counted.

FIGURE 4–20   Magnetic and photoelectric speed sensors

☐ TORQUE SENSING

Torque is the turning effort (twisting effect) required to rotate a mass or weight through an angle. Torque is a special type of work applied in a rotary manner around a center. It implies that the weight is located some distance from the center of the turning point. Torque is the force multiplied by the moment arm.

Torque is measured in inch-pounds (in/lb) or newton-meters (N-m). In some instances, it is measured in inch-ounces (in/oz). It can also be expressed in angular velocity. Angular velocity is the expression used when the torque is taken from a rotating motor shaft. These measurements are dealt with by mechanical engineers and robotic technicians.

Force and torque sensors are used to control robot motion. They may also make go/no-go decisions, adjust task and process variables, detect end effector collisions, determine required actions to unjam the end effector, and coordinate two or more arms.

One torque sensor system performs the above-mentioned tasks in robots. A basic component of the system is a six degree-of-freedom, solid-state piezoresistive force/torque transducer. The microcomputer automatically resolves the forces and torques applied to the transducer into six equivalent Cartesian force/torque components. It then transmits the results from the robot at better than 100 hertz rates.

The serial port of the microprocessor can be used to transmit force/torque information to the robot controller or to a readout device such as a cathode ray tube.

☐ VISION SENSORS

A robot's ability to see an object and make adjustments to fit that object are very important in certain types of manufacturing. Such an ability calls for some sophisticated sensing. The information obtained from a video camera has to be converted into digital signals before it can be processed by the computer and therein lies the problem.

Very intelligent controllers can make real-time adjustments in a robot's program in regard to its positional data. This is done to compensate for variables in the workpiece. This ability to compensate for variations in the workpiece is sometimes referred to as adaptive control.

To make these adjustments, the controller needs some sophisticated information about the current state of the process. This cannot be provided by the simple go/no-go limit switches and sensors generally used with nonintelligent controllers.

The type of information provided to the robot controller deals with where the part is located, how it is placed, and the type of part it is.

Machine vision systems (MVSs) are used for recognition and verification of parts. They can be used for inspection, sorting of parts, and for making noncontact measurements. MVSs also provide part position and orientation information to the robot controller. See Figure 4–21.

A number of imaging devices are used for robot vision systems. They include dissectors, flying spot scanners, videcons, orthicons, plumbicons, charge-coupled devices (CCDs), and others. These devices

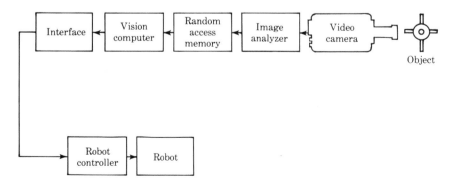

FIGURE 4–21   Machine vision system (MVS)

differ in the way in which they form images as well as in the properties of the images they form. One thing they all have in common is their ability to convert light energy to voltage in much the same way. Research is currently being done on a number of them. The most popular vision system has been the vidicon tube. It is the camera tube of a television system. However, the CCD system is becoming more popular and can be used in many industrial locations for robot vision. Fiber optics are presently being introduced into this field of sensing.

A detailed analysis of the various means of identifying parts for processing or inspection will be discussed in Chapter 7.

☐ SUMMARY

In order for the robot to hold and recognize objects, it must be able to sense their presence, size, and shape. Transducers are used to convert nonelectrical energy into electrical energy. A transducer then can serve as a sensor.

Limit switches are designed to be turned on and off by an object contacting a lever or roller that operates the switch. Some low and medium technology robots use this type of sensor.

Sensors are classified as contact or noncontact. They may also be called internal or external and passive or active. Most robotic sensors are contact or noncontact.

A limit switch is a contact sensor. Touch, force, pressure, temperature, and tactile sensors all respond to contact.

Pressure changes, temperature changes, and electromagnetic changes can all be sensed by noncontact methods. They usually react to a change in a magnetic field or light pattern.

A number of light-emitting diode (LED) sensors are used in robotics. The light beam is used in a number of applications. Another light sensor is the television camera mounted on the end of the manipulator. It can see the parts and compare them to what is in the memory of the computer.

The work envelope is that area where the manipulator moves the end effector. This area is usually enclosed by a cage to protect maintenance persons from being near the moving manipulator when it is operating.

Proximity sensors are used to give the robot a sense of touch and sight. They take various forms and work on different principles.

Different types of proximity sensors include the inductive sensor, capacitive sensor, and resistive sensor. Some have advantages that others do not. Pulsed infrared photoelectric controls are used in industrial robotics for presence sensing of any type of object. Eddy current proximity detectors use magnetism to function. They induce a magnetic field in an object, and a small coil is used to pick up the change in magnetic field. A reed switch is another magnetic electrical proximity switch. It responds to a controlled magnetic field.

One type of range sensor is called a laser interfero-metric gage. It is very expensive and is sensitive to humidity, temperature, and vibration. Another ranging system is the television camera.

Tactile sensors rely on touch. The simplest type is the microswitch. Pulsed infrared photoelectric systems can also be used for presence sensing. Some experimenting is being done to produce a better quality tactile sensor than is presently available.

Temperature sensing is done with thermocouples and thermistors. Displacement sensing is done with capacitive, inductive, and resistive devices.

The strain gage is a device used to sense mechanical movement and, in some cases, weight or force.

Speed sensing can be done with a tachometer or photocell. Torque sensing is used to make go/no-go decisions, adjust tasks and process variables, detect end effector collisions, and determine required actions to unjam the end effector, and to coordinate two or more arms.

Machine vision systems are used for recognition and verification of parts, as well as for inspection and orientation. Fiber optics are presently being introduced into this field of sensing.

☐ KEY TERMS

collision avoidance   the ability of a robot to avoid colliding with the part it is supposed to pick up

contact sensor   a sensor that detects the presence of an object by actually making contact

light-emitting diode (LED)   low-level light beam that is picked up by another device

limit switches   switches designed to be used with a moving body that should not go past a given point

machine vision systems   machine vision systems that allow the robot to recognize and verify parts

proximity sensors   devices used to detect how close an object is

range sensors   devices used to detect the precise distance between an object and the gripper or manipulator

strain gage   a device made of thin wire that reacts to stretching of that wire (Strain gages are made of semiconductor materials today.)

tactile sensors   devices used to detect the presence of an object by using touch

thermistors   devices that change their resistance in the reverse manner (If the temperature increases, the thermistor lowers its resistance.)

thermocouples   a union made by two dissimilar metals (When heated, the junction produces a small electrical current.)

transducer   a device used to convert nonelectrical energy to electrical energy

☐ QUESTIONS

1. What is a transducer?
2. How does a limit switch work?
3. List the two classes of sensors.
4. List the device used to sense touch, force, pressure, temperature, and vision in a robot.
5. What is an LED?
6. What is a work envelope?
7. What is meant by collision avoidance?
8. What does a proximity sensor do?
9. List three types of proximity sensors.
10. What are the two types of tactile sensors?
11. How many robots are used for a pick-and-place operation?
12. Describe how ACS is used in a touch sensor.
13. What are the two types of devices that sense temperature?
14. What is meant by displacement sensing?
15. How would you make a simple strain gage?
16. What do you use for speed sensing?
17. How is torque measured?
18. What is a machine vision system? Why is it needed?

CONTROL METHODS

Robots need a source of power to do work. The source of power may be one of three or any combination of three types of power, that is, electricity, hydraulic pressure, and pneumatic pressure. Hydraulic pressure and pneumatic pressure are driven by electric pumps or compressors. Let's take a closer look at these three sources of energy for the robot.

☐ ELECTRICAL POWER

Electric motors were discussed earlier in reference to the manipulator and the gripper. The DC brushless, DC permanent magnet, and the series, shunt, and compound wound DC motors also were covered. We have not, however, taken a close look at the single-phase and three-phase AC motors that power the manipulator and do a great deal of the heavy work. See Figure 5–1.

Motor Characteristics

	⊋	Duty	Typical Reversibility	Speed Character	Typical Start Torque*
POLYPHASE	A-C	Continuous	Rest/Rot.	Relatively Constant	175% & up
SPLIT PHASE SYNCH.	A-C	Continuous	Rest Only	Relatively Constant	125–200%
SPLIT PHASE Nonsynchronous	A-C	Continuous	Rest Only	Relatively Constant	175% & up
PSC Nonsynchronous High Slip	A-C	Continuous	Rest/Rot.†	Varying	175% & up
PSC Nonsynchronous Norm. Slip	A-C	Continuous	Rest/Rot.†	Relatively Constant	75–150%
PSC Reluctance Synch.	A-C	Continuous	Rest/Rot.†	Constant	125–200%
PSC Hysteresis Synch.	A-C	Continuous	Rest/Rot.†	Constant	125–200%
SHADED POLE	A-C	Continuous	Uni-Directional	Constant	75–150%
SERIES	A-C/ D-C	Int./Cont.	Uni-Directional●	Varying‡	175% & up
PERMANENT MAGNET	D-C	Continuous	Rest/Rot.§	Adjustable	175% & up
SHUNT	D-C	Continuous	Rest/Rot.	Adjustable	125–200%
COMPOUND	D-C	Continuous	Rest/Rot.	Adjustable	175% & up
SHELL ARM	D-C	Continuous	Rest/Rot.	Adjustable	175% & up
PRINTED CIRCUIT	D-C	Continuous	Rest/Rot.	Adjustable	175% & up
BRUSHLESS D-C	D-C	Continuous	Rest/Rot.	Adjustable	75–150%
D-C STEPPER	D-C	Continuous	Rest/Rot.	Adjustable	■

* Percentages are relative to full-load rated torque. Categorizations are general and apply to small motors.
■ Dependent upon load inertia and electronic driving circuitry.
● Usually unidirectional—can be manufactured bidirectional.

† Reversible while rotating under favorable conditions: generally when inertia of the driven load is not excessive.
‡ Can be adjusted, but varies with load.
§ Reversible down to 0°C after passing through rest.

FIGURE 5–1 Characteristics of electric motors (courtesy of BODINE ELECTRIC COMPANY)

Single-Phase Motors

Single-phase motors can be classified in a number of ways. Of primary concern here is the quick recognition of the general characteristics of the three single-phase motors (split-phase, capacitor-start, and shaded-pole) that are used in robots.

The split-phase induction motor is one form of the fractional-horsepower motor. This motor has two sets of windings: One is the run winding; the other is the start winding. The run winding is the main workhorse of this motor. The start winding is used only when the motor is started and is therefore called an auxiliary winding. Once the motor reaches 75 percent of its run speed, the start winding is taken out of the circuit by a centrifugal switch.

The split-phase motor has the constant-speed, variable-torque characteristics of the shunt DC motor. Most of the motors are designed to operate on 120 or 240 volts. For the lower voltage, stator coils are divided into two equal groups that are connected in parallel. For the higher voltage, the groups are connected in series. The starting torque is 150 to 200 percent of the full-load torque, and the starting current is six to eight times the full-load current. The direction of rotation of the split-phase motor can be reversed by interchanging the start winding leads. Fractional-horsepower split-phase motors are used in various machines and ventilating fans.

The capacitor-start motor is a modified form of the split-phase motor. It has a capacitor in series with the start winding. The capacitor produces a greater phase displacement of currents in the start and run windings than is produced in the split-phase motor.

There are three types of capacitor motors: the capacitor-start, the permanent-split capacitor, and the two-value capacitor. Each type has its own characteristics. One of the advantages of the capacitor-start motor is its ability to start under load and develop high starting torque. It can be reversed at rest or while rotating. The speed is relatively constant while the starting torque is 75 to 150 percent of rated torque. The starting current is normal. This type of motor can be used on air compressors and hydraulic pumps when the voltage rating is available. See Figure 5-2.

The shaded-pole motor comes in three types. It is built in a wide variety of ratings, with any number of performance characteristics. Shaded-pole motors are made in 0.00025 to 0.1 horsepower. Speeds for the shaded-pole motor are stated with no-load conditions. They are not very powerful. See Figure 5-3.

The shaded-pole motor has a low efficiency rating, a low power factor, a low starting torque, high noise and vibration, and is very low in cost. Having small poles shorted by heavy copper rings, it requires no starting mechanism. The shaded-pole motor finds applications in fans and small-horsepower devices such as clocks.

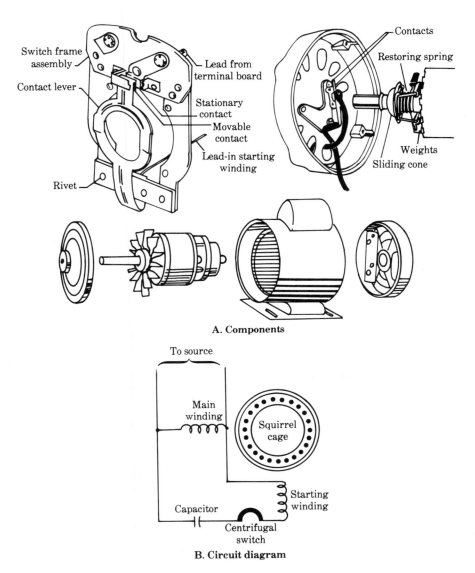

A. Components

To source

Main winding

Squirrel cage

Capacitor

Starting winding

Centrifugal switch

B. Circuit diagram

FIGURE 5-2 Capacitor-start single-phase motor (Note the large "bump" that houses the capacitor on top of the motor.)

Three-Phase Motors

Most industrial motors are three-phase, primarily because maintenance of a three-phase motor is practically nil. Industrial motors do not have the starting devices that single-phase motors have. The three phases of alternating current that supply power for the motor produce the phase shift needed to get the motor started and to keep it running once it is

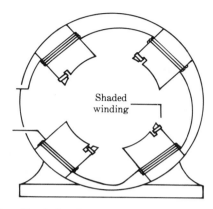

FIGURE 5-3   Shaded-pole motor

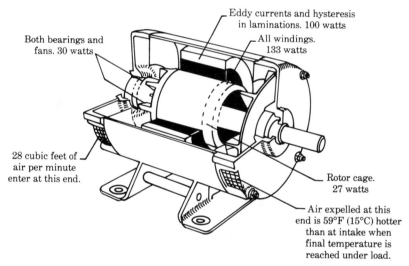

FIGURE 5-4   Three-phase electric motor

started. See Figure 5-4. All commercial power produced in the United States is generated as three-phase current. It can be converted to single-phase by dividing the three separate phases and sending them into three different subdivisions or locations. It is cheaper to distribute single-phase AC than three-phase AC. Three-phase power requires at least three, and sometimes four, wires for proper distribution.

Three-phase motors have good overall characteristics. They are ideal for driving machines in industrial uses. They can be reversed while running by reversing any two of the three connections to the power line.

The motor speed, under normal load conditions, is rarely more than 10 percent below synchronous speed. At the extreme of 100 percent slip, the rotor reactance is so high that the torque is low because of a low power factor. The three-phase motor has high starting and breakdown torque with smooth pull-up torque. It is very efficient to operate. It is available in 208–230/460 volt sizes, with horsepower ratings varying from one-quarter to the hundreds. It can be obtained for operation on 50 or 60 Hz.

If you classify robots by their power supply control, you get two different types of machines: servo-controlled and nonservo-controlled robots. A closer examination of these two types will produce some understanding of what is inside the large unit called the controller.

☐ SERVO-CONTROLLED ROBOTS

A servo-controlled robot is driven by servomechanisms. *Servo motors* are driven by signals rather than by a straight power line voltage and current. That means the motor's driving signal is a function of the difference between command position and/or rate and measured actual position and/or rate. The servo-controlled robot is capable of stopping at or moving through a practically unlimited number of points in executing a programmed trajectory. In other words, signals are produced that cause the robot to know where it is and where it is going. This is a more sophisticated system than that of the nonservo robot.

The servo-controlled robot can do more things than the nonservo type. It can move up and down and back and forth and is able to stop at any point within its work envelope. It can also return to the spot any time the program is run. This gives the robot many more uses than the simple bang-bang type. However, it is expensive. Some type of control system must be devised to make sure the robot can stop and go and reach a programmed point every time.

Feedback is the basic difference between nonservo and servo robots. *Feedback* is needed to tell where a manipulator is in reference to where it is supposed to be. This feedback system is what costs so much and produces so much research into methods of accomplishing it more effectively with less cost. See Figure 5–5.

To make the arm aware of where it is located within its work envelope, the controller must have some way of determining that the arm is not where it is supposed to be and then of making a comparison of where it is with information on where it should be. The controller sends out a command signal. A comparator compares the command signal from the controller and a feedback signal from the position sensor and registers an *error* in the position of the arm. This means the error has to be corrected. So the servo amplifier is fed the information, and it drives the hydraulic actuator to cause the arm to move to its programmed location. While the arm is moving, it feeds its positional information back to

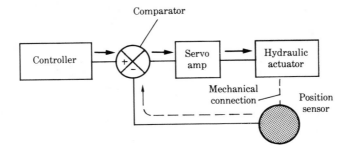

FIGURE 5-5 Feedback system

the comparator for comparison to see how close it is to correcting the error at any given point along the way. Once the error has been corrected and the arm is where it should be, the actuator stops supplying power to the arm.

□ NONSERVO-CONTROLLED ROBOTS

A nonservo robot is usually controlled by limit switches or by banging into stops at each side of its swing. It operates on very simple switching or limiting of its arm. Figure 5-6 is an example of a nonservo robot.

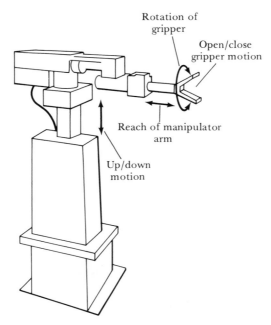

FIGURE 5-6 Nonservo-controlled, low technology robot (from Malcolm *Robotics*)

Sometimes referred to as bang-bang robots, nonservo robots are classified into three types, electric, hydraulic, and pneumatic, according to the means used to drive their manipulators.

Electric Nonservo Robots

This option for powering robots has been eliminated by virtue of the fact that electric motors do not take the stop and go associated with the swing of a robot arm. These motors are easily damaged by the constant and rapid changing of direction. Most nonservo robots use some other means of power.

Pneumatic Nonservo Robots

Air pressure can be used to power the manipulator arm of the nonservo robot. Air pressure is fed into the robot drive mechanism and the arm moves. Once the arm reaches the stop, it is at rest. In order to return to the other stop, it must be pushed back by air pressure from a second air line. The second line is activated and the first line is disconnected. A valve is used to disconnect one line and connect another. The valve is actuated by using an electrical current to energize a coil that draws the sliding bar or valve back and forth inside the unit. This works very well with the pneumatically operated robot since it can exhaust the air into the atmosphere every time it hits a stop. See Figure 5–7. (Actuators are discussed later in this chapter.)

Hydraulic Nonservo Robots

Hydraulic fluid under pressure can be used to power a nonservo robot. Since it is very hard on the robot to be constantly hitting the stop at full

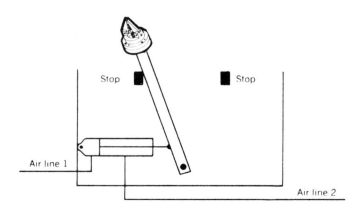

FIGURE 5–7   Pneumatic nonservo robot

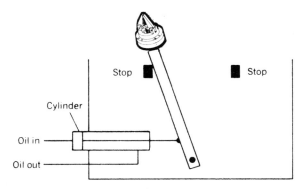

FIGURE 5-8   Hydraulic nonservo robot

speed, a switch is added to slow it down. A limit switch is located so that the arm will hit it just before hitting the stop. By opening the switch, the fluid in the hydraulic system is reduced. This allows the arm to coast to a stop against the stop. Shock absorbers also can be used to cushion the shock of sudden stops. See Figure 5-8.

□ ACTUATORS

Actuators are motors, cylinders, or other types of mechanisms used to power robots. They convert one type of energy to another. Actuators are employed primarily to provide the power to move each axis of the robot arm. Actuators are categorized by the type of motion they supply, either linear or rotary.

About 50 percent of the actuators used today are electrically driven. They may be stepper motors, DC servo motors, or pancake motors. Both small and large robots use electrohydraulic actuators. Electrohydraulic actuators make up about 35 percent of the devices used. Pneumatic actuators are the simplest and account for about 15 percent of the actuators in use. They are used in pick-and-place robots. See Figure 5-9.

Electric Actuators

The electric actuator is fast and accurate and easily adapted to sophisticated control techniques. It is simple to use and relatively inexpensive.

The electric actuator's disadvantages are its power limits, its backlash (since it requires gear trains for the transmission of power), and its tendency to arc, which may cause problems in an explosive atmosphere.

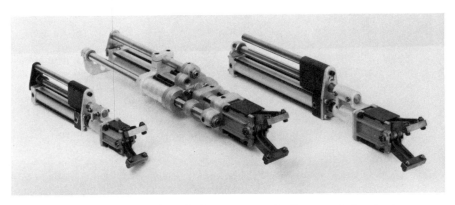

FIGURE 5-9   Pick-and-place pneumatically operated actuator (courtesy of I.S.I. MANUFACTURING, INC.)

Electric actuators are highly recommended where high accuracy, repeatability, and quiet operation are necessary.

Hydraulic Actuators

The hydraulic actuator is used when heavy loads must be lifted. It has moderate speed. Due to the characteristics of the hydraulic fluid, the joints once positioned, can be kept motionless, offering accurate control characteristics.

The disadvantages of the hydraulic actuator are its expense, its tendency to leak and make a lot of noise in operation, its slow speed in cycling, and its stiffness in operation.

Pneumatic Actuators

The pneumatic actuator is cheap, clean, and safe. Its lack of stiffness can be a disadvantage in some uses. It is high speed and does not pollute the workspace since it exhausts clean air into the atmosphere. See Figure 5-9. It is often used in laboratory work.

Some disadvantages of the pneumatic actuator are its noise and the problem of air leaks all along the system, especially when the air is under high pressure. This actuator requires good air filtering and drying and has a high level of maintenance and construction. It also has limitations due to the compressibility of air, which does away with its stiffness, and, in some cases, its joints creep instead of remaining motionless.

☐ CONTROLLERS

Six basic types of *controllers* in use today are the rotating drum, air logic, relay logic, programmable, microprocessor-based, and minicomputer. Each type of controller has advantages and disadvantages. In fact, some types are no longer widely used because of the rather rapid development of inexpensive electronic controllers.

Rotating Drum Controller

One of the first controllers to be used was the rotating drum. It was made up of a cylinder or mechanical drum that had cams on its surface. These cams would cause plungers to be activated and either release or press against them to open and close valves. These valves in turn controlled the air or hydraulic system to produce arm motion. The arm motion was halted by hitting a stop placed at the desired location. This type of controller had no feedback and no way of knowing where the arm was at any given time. The controller would repeat the same program every time the drum made a complete revolution. It is possible to design a program that takes into consideration the speed with which the drum rotates and the placement of the cams on the surface of the drum. The rotating drum resembles the workings of a music box with its rotating drum causing the various strings to vibrate.

Rotating drum controllers are used on pick-and-place robots and are very reliable. They do require maintenance from time to time, but they can continue to perform without attention for long periods. Thousands of them are still in operation.

Air Logic Controller

Air logic controllers were used on pick-and-place robots in the sixties and seventies. They are very useful in explosive atmospheres since they do not have electrical switching to create sparks. A series of air valves are placed so that the arm is actuated by the valves at the end of a cylinder's desired travel. The valves are connected together with short pieces of tubing that supply the air pressure needed for operation. In other words, a valve closes, creating pressure that causes the arm to move and then opens, releasing the pressure, which lets out the air. Another valve closes and the air pressure causes it to move again. By placing the valves where needed, it is possible to push the arm back and forth over a path desired for its operation.

Relay Logic Controller

Relay logic controllers were used with machines before robots were developed. An electromechanical switch, called a relay, was used to make

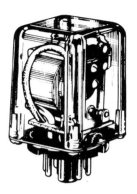

FIGURE 5-10   A relay

sure switches were closed in the right order to perform a task. See Figure 5-10. A *ladder diagram* was developed for each operation. See Figure 5-11. The circuit (ladder diagram) is nothing more than the two lines coming from the power supply. Various relays are connected between these lines. Limit switches and other actuating switches that are mechanically closed or opened by the robot's manipulator or cylinders cause the relays to open or close to perform the required operation. The *relays* cause the *solenoids* to be energized and to open and close the pneumatic or hydraulic valves. The valves make the robot arm do what it is supposed to do. Ladder diagrams are used today for programmable controllers and other electrical equipment.

Relay switches need a great deal of attention and cause a lot of downtime for the machine. Industry has quickly replaced the relay with electronic components, just as the telephone company now uses transistor switching instead of relay switching. There are no moving parts in transistors and no contact points to be cleaned periodically. With relays, the whole circuit had to be rewired to make simple changes. In semiconductor devices, no rewiring is necessary; one needs only to remove a board and substitute another or, in most instances, reprogram without removal.

The rotating drum, air logic, and relay logic controllers are not used today for robots. Less expensive controllers with no moving parts have replaced the maintenance-prone types. The older types had to be reprogrammed along with a lot of physical work. They could not communicate with other robots or with a central location.

Programmable Controller

The programmable controller (PC) is the next step up in controllers. See Figure 5-12. It is a little more sophisticated than the drum, air logic, and relay logic types.

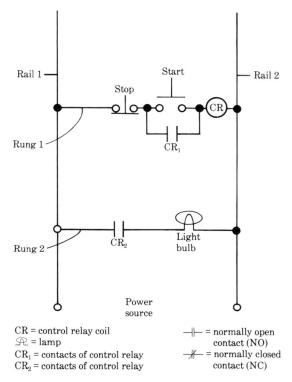

CR = control relay coil —||— = normally open
𝓡 = lamp   contact (NO)
CR₁ = contacts of control relay —|/|— = normally closed
CR₂ = contacts of control relay   contact (NC)

Push start button and complete the circuit to CR coil. Coil
energizes and closes CR_1 and CR_2. This causes CR to
remain energized until stop button is pressed to open the
circuit. When the relay is energized, CR_2 contacts are also
closed, causing the lamp to light and show power on.

FIGURE 5–11   Ladder diagram

The programmable controller uses electronics for timing and sequencing. Instead of using pegs to store the memory as the drum controller did, the programmable controller has an electronic circuit for
memory. Many programmable controllers are programmed from a keyboard that is similar to a typewriter keyboard. The keyboard is used to
enter the proper order of switch closings.

The programmable controller is a computer-based device that is
nothing more than the electronic way of doing what the relays did. The
main advantage is that the PC program can be changed quickly and easily without physical work. Maintenance is less expensive, and the cost
has become very low in comparison to previous controllers.

PCs are used in a variety of applications to replace conventional
control devices such as relays and solid-state logic. When compared
with conventional controllers, PCs allow ease of installation, quick and
efficient system modifications, more functional capability, troubleshoot-

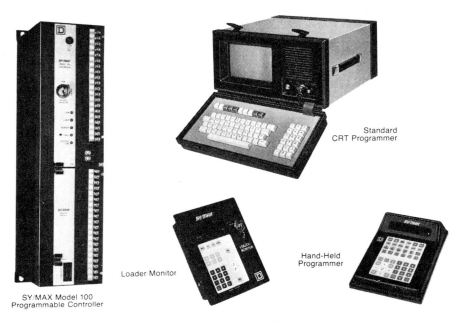

Standard
CRT Programmer

Hand-Held
Programmer

Loader/Monitor

SY/MAX Model 100
Programmable Controller

FIGURE 5-12   Programmable controller programming equipment (courtesy of SQUARE D COMPANY)

ing diagnostics, and a high degree of reliability. Typical installations include automated materials handling, machine tooling, assembly machine control, wood and paper processing control, injection molding machine control, and process control applications such as film, chemical, food, and petroleum.

The PC consists of system hardware and programming equipment. System hardware can be a processor, one or more rack assemblies, power supplies, input/output (I/O) modules, and various other modules that provide additional capabilities. The rack assemblies and associated I/O modules communicate with external I/O control devices such as limit switches and motor starters. System hardware may also be a PC that incorporates all necessary hardware in a single package.

Programming equipment consists of either a hand-held programmer or a deluxe CRT programmer that is used during program entry or for monitoring the operation of the system. A separate loader/monitor for monitoring and changing data or printing messages can also be obtained.

PC processors provide relay logic, latch/unlatch relays, data transfers, timers, counters, master control relays, synchronous shift registers, data comparisons, bit read and control, transitional output, and register fencing; respond to communications from other processors; and have I/O forcing capability.

As you can see from this terminology, a number of new terms apply to this rapidly advancing field. It is difficult to do justice to the topic of programmable controllers within the confines of this chapter. An entire book is needed to develop the technical expertise associated with PC devices and their capabilities. However, this brief introduction to PCs indicates how they are connected to the advancement of the robot and the field of robotics.

Microprocessor-Based Controller

The *microprocessor* is the building block of the microcomputer. Microprocessors are designed to do a specific job. In this book, we are concerned with how a microprocessor is designed to work with a robot and give it certain capabilities to perform tasks. See Figure 5–13.

Microcomputers are used with all types of robots. The microprocessor is specially designed for a specific robot. It has a memory and the circuits needed to cause the robot to function and be easily programmed or edited. In some cases, the microcomputer will have a cathode ray

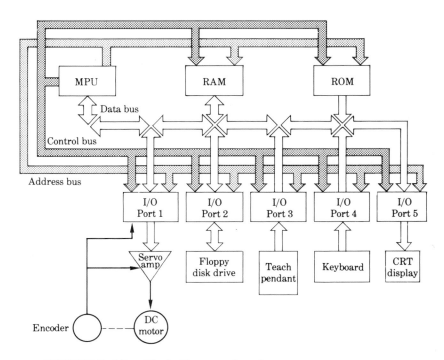

FIGURE 5–13   Block diagram of a microprocessor

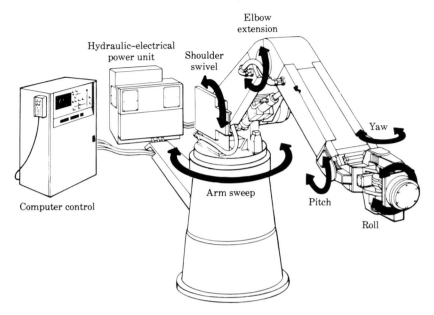

FIGURE 5-14   Robot system with minicomputer (courtesy of
CINCINNATI MILACRON)

tube (CRT) to display the contents of its memory and the program that
has been placed in it.

The output of the program must be able to power the solenoids
that control the manipulator arm. This calls for an amplifier to take the
very weak signals from the computer and boost them sufficiently to energize solenoids that control hydraulic or pneumatic valves.

Minicomputer Controller

Minicomputers are used with some robots to give them a greater level of
sophistication and to enable them to do tasks that could not be done
otherwise. Robot manufacturers have mated the computer with their robots to expand their applications. The output of the computer is fed into
circuits to be amplified and then fed to the control devices for the manipulator, wrist, and gripper. See Figure 5-14.

☐ PROGRAMMING THE ROBOT

Robots have to be "taught" to do jobs. They are easily programmed or
taught, once their electronics packages are understood. Electronics
makes it possible for the robot to do so much.

Putting pegs in a drum and changing air hoses as a means of programming a robot have already been discussed. This technology is obsolete. Now, besides keyboarding as a way to teach a robot a task, the teach pendant method, lead-through programming, and computer terminal programming are commonly used.

Teach Pendant

The *teach pendant* is the most popular device for programming a robot since it requires very little mathematical figuring and editing of a program. The desired position for the manipulator is reached by walking it to the spot where it is needed. This walking is done by pressing buttons on the teach pendant that cause the arm to move where you want it. A teach pendant button is then pressed to store the information in memory. See Figure 5-15. The manipulator is led to the next desired position by pressing the right buttons. Once it is in position, that too is entered into the memory of the robot as the next point in the program. None of the moves made by the programmer between the first point and second point is remembered by the robot; the robot remembers only the final points. The program is played back and the robot's manipulator moves smoothly from the first point to the next point stored in its memory.

The program is tested before the robot is put into production

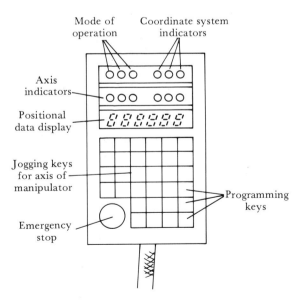

FIGURE 5-15   Teach pendant (from Malcolm *Robotics*)

runs. The speed from one point to the other can also be programmed with the teach pendant. This allows the operator to make sure the arm moves smoothly and comes to a stop without abrupt action. It is also possible to program the acceleration and deceleration of the arm by using the pendant. Smoother operation is then possible. The programmer steps through the program after it has been stored, recalling all the information and putting the manipulator through its motions. If errors are detected, they can be edited out with the teach pendant before the robot is put into operation on the line.

Once the program has been checked for correctness, the robot can be put to work, and it will repeat the motions for as long as needed. You will see why this is the easiest way to program the robot once you have looked at other methods of programming.

Lead-Through Programming

Lead-through programming is usually done with continuous-path robots. They can be programmed to follow a circle, an arc, or a straight line. This type of robot is programmed by grabbing the arm and moving it through the actions needed to do the job. The robot will memorize the path and then play it back when called for. The speed at which the arm moves can be adjusted later after the exact strokes or movements are committed to memory. Adjustments can also be made for small errors that are detected before it is put into production. This type of robot is useful as a painter or welder. For example, an experienced painter leads the arm through the same motions he would make if he were doing the job. The robot memorizes the path and then repeats it over and over again. This type of robot can remember thousands of points between start and stop. Figure 5-16A shows a spray painting robot that is easily programmed by using the lead-through teach method. Figure 5-16B is an editing module used to correct errors in speed and movement.

Computer Terminal Programming

Using a computer to program a robot can be difficult. Robots use a number of languages to move the manipulator from one place to another. These languages can be used to call up the program that was memorized by a walk-through and modify it for smoother and more precise operation. A computer language can also be used to write a program from scratch. The computer program comes in handy when the robot is to work in concert with a welder, conveyor belt, or other types of equipment that make up a work cell. See Figure 5-17.

Computer controllers were developed in the 1960s when integrated circuits (ICs) were introduced. The integrated circuit is a complete circuit with its resistors and transistors on a single silicon chip.

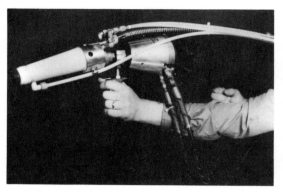

A. Spray painting robot that is easily programmed using the lead-through teach method

B. Remote editing module that makes small changes in programming during each operating cycle

FIGURE 5–16   Lead-through programming and remote editing (courtesy of BINKS MANUFACTURING COMPANY)

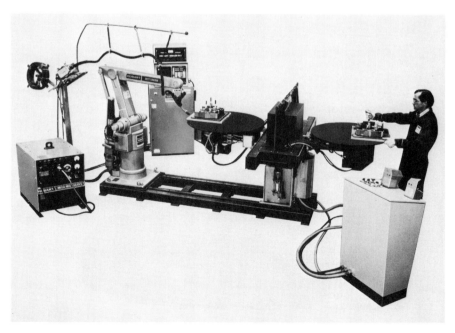

FIGURE 5–17   Work cell with robot controller and welder (courtesy of HOBART BROTHERS COMPANY)

See Figure 5-18. These *chips* may have hundreds of transistors and diodes in a single package, which make possible microcomputers in very small packages. The chips also allow for the design of a computer for each robot. The price of electronic components has decreased so much that it is now cheaper to buy an electronic controller than it is to buy the drum or air logic controllers. The extended capabilities of computer terminal programming are also much more economical when it comes to keeping the robot operational.

Inexpensive computer terminal programmers make it possible to have vision, touch, and hierarchical control for almost all robots.

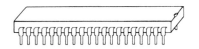

**A. 8088 microprocessor, a popular
16-bit microprocessor chip**

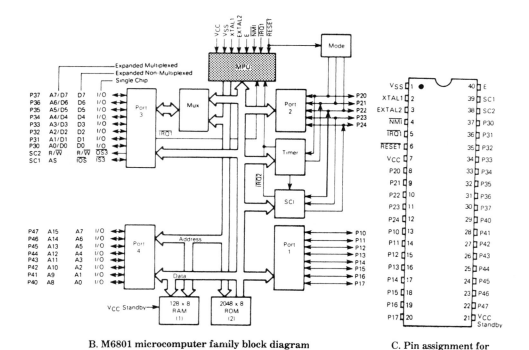

B. M6801 microcomputer family block diagram

**C. Pin assignment for
the MC6801**

FIGURE 5-18 Integrated circuit chip (courtesy of MOTOROLA, INC.)

Desktop computers such as the Apple can be used to program a robot. The Apple communicates with the robot through the robot's controller. The controller actually operates the robot with the information furnished by the computer. The controller has a microprocessor that has to be *interfaced* (properly connected) with the computer.

□ SUMMARY

Robots need a source of power to do work. The power may be from a single source or from any combination of three sources, that is, electricity, hydraulic pressure, or pneumatic pressure.

Single-phase and three-phase motors are used to provide the energy to move heavy loads. Single-phase motors may be either split-phase, capacitor-start, or shaded-pole. The shaded-pole motor is usually employed to power fans and ventilation devices. The split-phase motor does not start well under load, but the capacitor-start motor does. The capacitor-start motor can be used to power compressors and similar devices where lower voltages (120/240) are available. The three-phase motor is one of the most reliable of electric machines. It is the workhorse of industry.

A servo-controlled robot can do more things than a nonservo-controlled robot. It can move up and down and back and forth and is able to stop at any point within its work envelope.

The nonservo robot usually is controlled by limit switches or banging into stops at the end of each swing. There are electric, pneumatic, and hydraulic nonservo-controlled robots.

Feedback is the main advantage that servo-controlled robots have over nonservo-controlled robots. Feedback tells the controller where the manipulator is located at all times.

Actuators are motors, cylinders, or other mechanisms used to power robots. They are employed primarily to provide the power to move each axis of the robot arm. The actuator causes the motion of the robot. There are pneumatic actuators, electrically operated actuators, and electrohydraulic actuators.

Controllers are available in six types: drum, air logic, relay logic, programmable, microprocessor-based, and minicomputer. The drum, air logic, and relay logic controllers have become obsolete with the advent of the integrated circuit chip and its ability to store and recall programs for the robot.

The ladder diagram is the circuit used by the programmable robot. It is also needed to make the computer function as a device that can control sequencing and timing of operations of a robot. The computer has replaced the switching operations normally done by a relay. The relay turned on and off the solenoid that allowed air or hydraulic fluid to pass or exhaust. The ladder diagram is the electrical schematic of the circuit of control for the solenoids and the timers.

Microprocessor-based controllers were made possible by the development of the integrated circuit chip. They have the ability to store sequences and allow them to be recalled when needed. This opened up the possibility of making changes to a program without having to mechanically adjust the circuitry.

The teach pendant is one of three ways to program a robot. It is used for point-to-point programming of pick-and-place robots and for programming continuous-path robots used for painting and welding.

Lead-through programming is done by leading the manipulator through the points it is supposed to follow. The points are stored in its memory and recalled whenever the program is repeated.

Computer terminal programming is done with the aid of a computer properly connected to the controller of the robot. This then allows for easy changes in the program if needed to change the job of the robot.

☐ KEY TERMS

actuators   motors, cylinders, or other mechanisms that are used to power robots

chips   integrated circuits that contain many transistors, diodes, and resistors to make up various electronic circuits on a small (about 8 mm square) piece of silicon; components that store the computer's program and act as its memory

controllers   units needed to control the robot by preparing the proper signals and timing to cause the robot to operate with some degree of accuracy and repeatability

feedback   ability of a device to feed a signal back to its controller to aid in keeping track of the position of the manipulator or gripper

interface   proper connections between a robot and its computer or microprocessor

ladder diagram   electrical circuit drawing of the sequence of switches needed to cause a robot to perform programmed activities

lead-through programming   taking the continuous path robot manipulator and leading it through a sequence of motions to accomplish a task

microcomputers   small-scale computers composed of microprocessors

microprocessor   brain of a microcomputer; an integrated circuit (IC) that performs arithmetic, logic, and control operations within the microcomputer

minicomputers   mid-size computers, slightly larger than microcomputers and smaller than mainframes

relays   electromechanical devices that are energized and that close or open switches

servo motors   motors driven by signals rather than by straight power line voltage and current; motors whose driving signal is a function of the difference between command position and/or rate and measured actual position and/or rate

solenoids   coils of wire with plungers that can turn on or off a fluid or air line

teach pendant   device used to teach the robot memory a new program

☐ QUESTIONS

1. What type of power does the power supply have to furnish the robot?

2. Name three types of single-phase electric motors.

3. Which type of single-phase motor has good starting torque?

4. Name three types of nonservo-controlled robots.

5. What is the advantage of the servo- over the nonservo-controlled robot?

6. What is the basic difference between the nonservo- and servo-controlled robot?

7. What does an actuator do?

8. How are the greatest number of actuators driven today?

9. Name six types of controllers in use today.

10. Why have the drum and air logic controllers become obsolete?

11. What is a ladder diagram?

12. Why are relays used in a ladder diagram?

13. How is a programmable controller programmed?

14. Describe the microprocessor controller.

15. What is a teach pendant?

16. How do you program a pick-and-place, point-to-point robot?

17. How do you use a teach pendant to program a welder robot?

18. How is lead-through programming done?

19. What is an IC chip?

20. What does the term *interface* mean?

THE ROBOT AND THE COMPUTER

C onstant supervision is necessary to ensure that robots do
their assigned tasks the way they are supposed to do them.
Many robot manufacturers are now building in this supervision
capability. The controller is built into the control system, and it
works much the same as the programmable controller. Inasmuch
as the robot is always interacting with conveyors, transfer lines,
and other materials-handling equipment, it is important that the
robot be able to function under these conditions. To make the
robot a safe machine, it needs switches interconnected to the
other equipment to stop it in case a dangerous situation arises.
All this inputs to the controller or microprocessor.

☐ ROBOT–COMPUTER INTERFACE

A number of methods or systems are used to accomplish the mating of computer and robot. Figure 6-1 is a schematic diagram of the microcomputer system that is often used for this purpose. As shown, the three major parts of the microcomputer are: memory, the central processing unit (CPU), and input/output (I/O) ports.

Memory

The computer's memory is used to store data to be processed by the CPU or data resulting from processing.

Central Processing Unit

The *CPU*, sometimes called the microprocessing unit (*MPU*), contains the control circuitry, an arithmetic logic unit (ALU), registers, and an

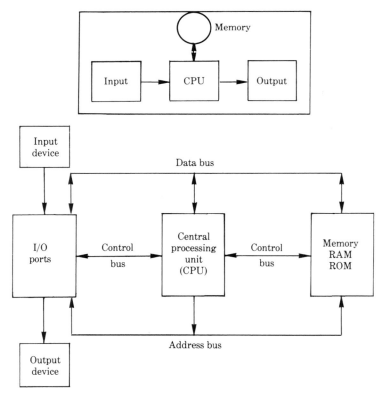

FIGURE 6-1   Block diagram for a robot computer system

address program counter. This may be a larger computer in larger installations.

The instruction code comes to the CPU from memory and gets decoded and executed. The CPU works in millionths of a second, enabling it to recall an instruction and process it without a slowdown in operation. Instructions are given in many computer languages. Each manufacturer specifies which language or languages the machine will use.

Input/Output

The connections that interface with the outside world of the microprocessor are called input/output ports. The input port allows data from a keyboard or other input device to be taken into the CPU. The output port is used to send data to an output device such as a motor. Bus lines carry the signals to and from the major parts of the microprocessor.

☐ LANGUAGES

Intelligent robot controllers have the ability to understand high-level languages. Almost every robot manufacturer has its own controller *language*, making the language unique for a particular controller from a particular manufacturer. Robots are usually preprogrammed for the benefit of the user. Some of the more frequently used languages are: VAL, HELP, AML, MCL, RPL, and RAIL. These are relatively simple languages. BASIC, COBOL, and other languages can also be used.

VAL was developed for Unimation. HELP was developed for General Electric's assembly robots. AML, which is a manipulator language, was developed for IBM's assembly robot. MCL stands for manufacturing control language and was developed by McDonnell Douglas for an Air Force project. Cincinnati Milacron's T^3 robot uses it. RPL stands for robot programming language and was developed by SRI. It is similar to languages such as Pascal and FORTRAN and is useful for communicating with intelligent vision sensors. RAIL was developed by Automatix for robots and vision systems.

Following is an example of a program written in VAL that will give you an idea of how a robot language works.

VAL

The robotic language VAL was developed by Unimation for the specific purpose of controlling the robots that they manufacture. VAL is also used by Adept and Kawasaki. Since Unimation robots are widely used, Rhino chose to emulate the Unimation language, VAL, for use with the

XR Series robotic system. Following is a Unimation program written in VAL and used to palletize products:

```
 1.              SETI      PX = 1
 2.              SETI      PY = 1
 3.      10      GOSUB      100
 4.              IF PX = 3      THEN 20
 5.              SHIFT PALLET BY 100.0, 0, 0
 6.              GOTO 10
 7.      20      IF PY = 3      THEN 40
 8.              SETI PX = 1
 9.              SETI PY = PY + 1
10.              SHIFT PALLET BY −900.0, 100.0, 0
11.              GOTO 10
12.      100     APPRO CON, 50
13.              WAIT CONRDY
14.              MOVES CON
15.              GRASP 25
16.              DEPART 50
17.              MOVE PALLET:APP
18.              MOVES PALLET
19.              OPENI
20.              DEPART 50
21.              SIGNAL GOCON
22.              SETI PX = PX + 1
23.              RETURN
24.      40      STOP
```

This VAL program causes the robot to go through the following procedures to pick up an object from a conveyor belt and deposit it on a pallet:

- Lines 1 and 2 keep track of how many parts have been loaded onto the pallet in both the x and y directions.
- Line 3 calls a subroutine (GOSUB) that will unload one part from the conveyor and place it on the pallet.
- Line 5 redefines a coordinate frame named PALLET. SHIFT on line 5 implements a translation of $x = 100.0$ mm, $y = 0$, $z = 0$.
- Inside the subroutine, the robot approaches the frame named CON.
- The APPRO function causes motion to a point displaced 50 mm out the z axis of the named frame and oriented such that the z axis of the hand is aligned with the z axis of the named frame. The APPRO command also specifies joint interpolated control.

- Line 13 tells the robot to wait for a signal (WAIT CON-RDY) from an outside source such as a limit switch. The limit switch indicates that the conveyor is ready.
- Line 14 moves the manipulator to grasp position and closes the gripper. MOVES indicates Cartesian straight-line motion.
- Line 15 halts the program if the hand closes to less than the minimum anticipated distance of 25 mm.
- Line 16 specifies Cartesian motion along the z axis of the hand frame to the departure point. The 50 indicates 50 mm out.
- Line 17 shows the alternate form used in VAL. It could have been APPRO PALLET, 50. MOVE specifies joint interpolated motion. (Note the difference between MOVES and MOVE.)
- Line 21 indicates how VAL can output signals to other devices. In this case, the command SIGNAL GOCON starts the conveyor in motion.

Unless a person is well versed in computer languages and the programming of computers, it is almost meaningless to go through a program in any greater detail than has been done here. It is significant to note, however, that a more thorough knowledge of programming is needed in order to become a robotics technician.

☐ SOFTWARE

Computers are used to program robots, and the programming is done *off-line*. That means the robot is not directly involved when programming is taking place. The *software* can be recorded on floppy disks, magnetic tape, or other means and then checked on a robot and debugged or edited. Once the program is proven on a robot system, it can be duplicated and sold to others who want to use robots to perform similar tasks.

By using off-line programming, the programmer has greater flexibility to carry out complex operations, and the time spent in programming is reduced. In addition, the robot can remain in service while the programming is taking place, thereby increasing its productivity.

☐ INTERFACING

Interfacing links permit the robot to communicate with its controller and other parts of the work station. Parts that do not cause the robot to

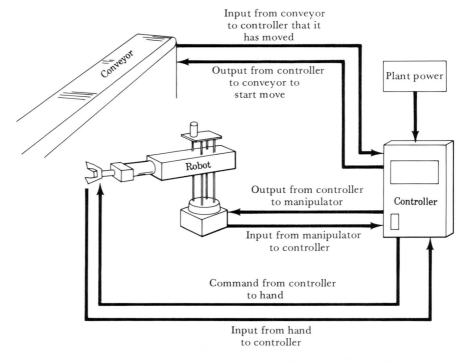

Input from conveyor
to controller that it
has moved

Output from controller
to conveyor to
start move

Plant power

Output from controller
to manipulator

Controller

Input from manipulator
to controller

Command from controller
to hand

Input from hand
to controller

FIGURE 6-2   An interfacing link (from Malcolm *Robotics*)

do its task directly are called peripheral components. Figure 6–2 shows a simple interfacing link.

The microprocessor-controlled robot can communicate with the equipment around it through connections called *ports*. For example, it is necessary for the robot to input information to the controller, and it must be able to receive information from the controller. Hence, the controller is connected to the robot through a port.

□ ASCII CODE

In order to communicate with the controller, you must be able to input instructions to its microprocessor. A special code has been designed that allows a regular typewriter keyboard to be used to type in instructions. However, the keys of the keyboard are actually switches that send electronic pulses to a decoder, which generates a special binary code. The code most often used is *ASCII*, or American Standard Code for Information Interchange.

The ASCII code is made up of seven binary bits, with 128 possible combinations (obtained when you take 2 to the 7th power, 2^7). The 128, or 2^7, possible combinations of 1's and 0's represent all the letters of the alphabet, both upper and lower case, as well as the numbers 0 through 9 and several special codes that include punctuation and machine control information.

Parallel Ports

Parallel ports are the outputs of the microprocessor or computer that have flat cables connected to them with eight conductors. Seven of these wires or conductors carry the information mentioned above. The eighth conductor carries the *strobe line*, the one that prevents switch bounce. When a switch is closed, it bounces and can allow more than the on and off information to be given. This noise or incorrect signal information must be prevented from being transmitted from the keyboard to the controller and then to the robot. It could cause the robot to make an incorrect move.

Serial Ports

The serial format may also be used to transmit data in the ASCII code. Serial ports allow the information to be transmitted along two wires. This makes transmitting over long distances feasible. The parallel format is very good for short distances or connection between machines within the same work cell, but if the information has to be sent for a greater distance, it is better to send it by serial formatting.

The information may be transmitted as changes in voltage or changes in current. There are standards for both. In fact, there are two standards for each. The two voltage standards are known as RS-232C and TTL. The two current standards are the 60 mA current loop and the 20 mA current loop.

The *RS-232C standard* says that the voltage of the signal will be between -3 and -25 to represent the logic 1 or "on" condition. A voltage between $+3$ and $+25$ is used to represent the logic 0 or "off" position. This standard was developed by the Electronics Industry Association (EIA). The advantage of this standard is that the line noise will have to be very high to make any false signals, and the voltage losses along the line will not affect the signal level as much as lower voltages do. Its disadvantage is that it has to be converted to TTL at the port of the computer.

The *TTL standard* is compatible with transistor–transistor logic and interfaces directly. There are problems with any transmission of data over a distance. If there is a line voltage of at least 0.5 volt, then there is the possibility of receiving incorrect data. Since the peak is only 5 volts, there is always the possibility of picking up a noise signal when a wire is spread over a distance in an electrical noise-generating environment.

The *60 mA standard* says that a current of 60 mA is logic 1, whereas zero current represents logic 0. The main advantage is that the noise usually encountered over long distance transmission lines does not affect the quality of the data being transmitted. However, the main disadvantage of this standard is that it has to be converted to voltage variations if used as inputs to a computer port.

The *20 mA standard* is basically the same as the 60 mA standard except that it is 20 mA. The 20 mA level represents logic 1, and zero current represents logic 0. The same advantage is experienced with this standard as with the 60 mA one. It is also necessary to convert the current variations to voltage variations if used as inputs to a computer port.

□ INTERFACING ROBOT AND COMPUTER

The controller has input ports for interfacing with various computer controls. RS-232C and RS-422 formats are used for this interfacing. RS-422 is an improvement of RS-232C. A serial format is used for inputting. The information fed to the computer has a start bit, a parity bit, and a stop bit, as well as the data information. Full duplexing is used, which means that the port for the computer interface can send and receive data using the same line.

A computer interface allows the programmer of the robotic system to program off-line. That means the service codes, geometric moves, and velocity rates can be programmed off-line. They are then downloaded to the controller.

The RS-232C can send information for about 50 feet maximum. An amplifier is needed to ensure that a lot of noise and incorrect information is not picked up. The amplifier boosts the signal sufficiently to keep the signal voltage up to the level required to overcome noise. Noise is everywhere in an industrial plant. Motors turning on and off generate spikes that can be interpreted as data pulses.

Sensors

Sensor information is converted to digital code so the computer can handle it. Controllers have an interfacing port that provides for connecting sensors. The RS-232C port is usually used with full duplexing. Information from the sensors is sent to the controller on its data bus and stored in internal memory.

☐ PROGRAM CONTROL

Program control of the periods when data are transmitted and received by the interfacing ports is important to the operation of peripheral components. The interfacing is such that program control of the sequence of events becomes very important if the robotic cell is to function properly.

There are two types of program interfacing: *service requests* and *robot requests*. Service requests deal with the interfacing operation for peripheral components and provide program control of the periods when the data are transmitted and received at the input ports. See Figure 6-3.

In order to control the sequence of events in a work cell, the person doing the programming must take into consideration when the signals will be outputted from the controller and what types of signals will be needed by the work cell.

Robot requests also require signals for operation. See Figure 6-4. The input and output signals come from either the MPU or the controller. In this case, the controller has been programmed for the sequence of operations that the robot has to perform to open and close the gripper and do other tasks. The controller must also have some method of finding out if the gripper actually picked up the object intended. As you can

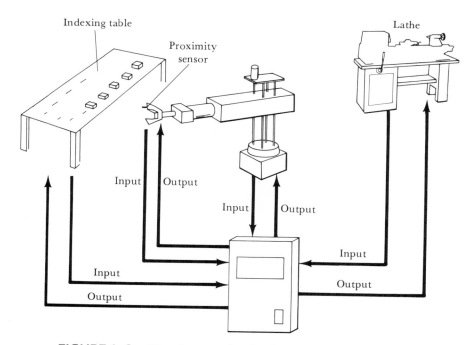

FIGURE 6-3   Signal processing for the operation of a work cell (from Malcolm *Robotics*)

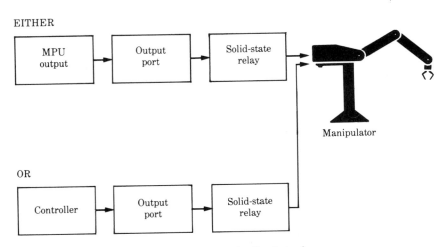

FIGURE 6-4   MPU output or controller interface

see, the input and output signals keep the robot working. The signals also provide communication between and among components in the work cell. This is very important in any automated production line. The robot has to work in sequence with the conveyor belt and do its job at the right moment. Timing is very important in any group effort. This is where the computer is at its best. It has the ability to organize many operations in a timely fashion and then to check to see whether they are done in sequence. The computer is also capable of keeping track of vast amounts of information for both the robot and the peripheral components.

□ VISION FOR THE ROBOT

Providing the robot with vision is one of the major challenges today. A number of different approaches have been tried, and some degree of success has been achieved with several of them. In fact, the auto industry is using vision inspection for a few tasks. Very intelligent controllers have the ability to make real-time adjustments in a program's positional data. This is done to compensate for variables in the workpiece. This ability is sometimes referred to as adaptive control. To make adjustments, the controller needs more sophisticated information about the current state of the process than can be provided by go/no-go limit switches and sensors used by nonintelligent controllers.

Vision systems provide the robot controller with information about the location, orientation, and type of part to be handled. A machine vision system (*MVS*) may be used for recognition and verification of parts, for inspection and sorting of parts, for noncontact measurements, and for providing part position and orientation information to the robot controller. See Figure 6–5.

The video camera can be used to take photographs, and the human brain can react to variations in shades of gray in those photos. See Figure 6–6. However, in order for a computer to be able to react to the information, it must be converted to digital format. That means some method of shade interpretation must be used. A gray scale is developed, and its parts are given numbers that correspond to digitization. The computer can then handle the information but must be programmed in

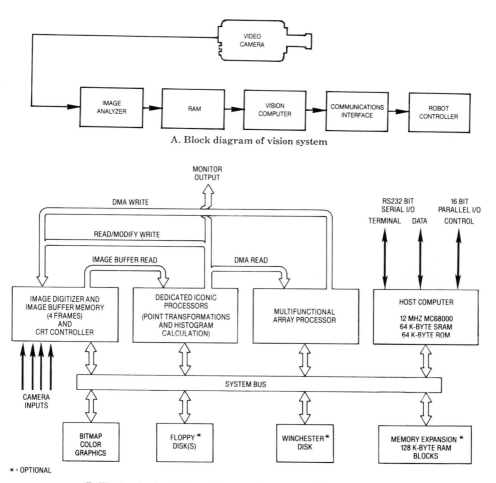

A. Block diagram of vision system

B. Electronics for taking and presenting camera signals and processing information to a monitor and computer

FIGURE 6–5   Machine vision system for robots (part B courtesy of INTERNATIONAL ROBOMATION/INTELLIGENCE)

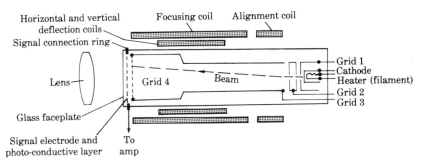

FIGURE 6–6   Video tube used in a robot vision system

some acceptable fashion to be able to compare what it is looking for and what is in front of it. Here a problem arises: There is no standardized method for making machine vision. A number of different approaches have been tried and are available on the market for use in limited processing operations.

Object Recognition

Object recognition is the main reason for putting vision on the robot arm, that is, to make the robot capable of locating and distinguishing objects or parts. Images of the objects to be recognized are identified by the video camera. The information is then converted to digital signals and stored in the computer's memory. This is the training phase of the operation. Then the robot, through the vision sensor, distinguishes one part from many different parts grouped together. There are two methods for distinguishing an object: the edge detection process and the clustering process.

Edge detection uses the difference between light and dark areas of an object to make distinctions. See Figure 6–7A. A camera, located on the end of the manipulator arm or somewhere along the conveyor belt, locates an object – in this case, an automobile window. The image of the object is broken down into digital pulses. The pulses are sent to the vision computer. The vision computer searches through its memory and compares the different gray areas of the image with the gray scale already stored in its memory. When the vision computer finds a match between areas that contain the correct gray scale variations, it has located the edge of the part.

Figure 6–7B shows how the information from the video camera is converted to digital information in an analog-to-digital converter before it is fed to the computer. Note that the pixel clock generates a series of pulses to be fed into the analog-to-digital converter, with the video providing the gray level in digital form.

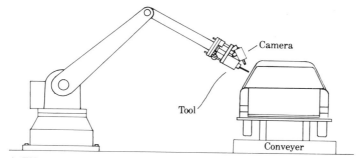

A. TV camera mounted on robot for visual inspection on auto assembly line

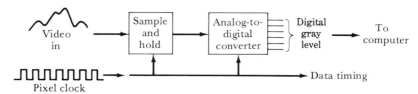

B. Pixel clock and data bit generation for a video signal fed to a computer

FIGURE 6-7   Example of the edge detection process (part B from Malcolm *Robotics*)

The *clustering process* is similar to the edge detection process. It processes the camera signal and feeds it to the computer. The computer compares the various contrast levels to what it has stored in memory, and if they match, the object has been recognized. Both the edge and the clustering processes have a problem with signal interference. In any industrial environment, there are a lot of voltage spikes (noise) generated by the motors and other equipment being cycled on and off. These voltage spikes produce pulses that can be misinterpreted by the computer. The cluster and edge detection processes both require a high light level for the camera to work properly and to be able to pick up the contrast in object shapes.

The template method of part recognition shows some merit. It takes an image and rotates it until it matches the template stored in the computer's memory. The computer must have a large memory capacity to be able to handle the rotation and comparison process. See Figure 6-8.

Programs can be written to aid the computer in doing its job. The programs can also be written to cause the robot to reject a part or put it in a different bin or container than that of the desired object.

Much work is being done on robot vision systems. However, there is much to be done before robots have the ability to recognize objects and identify them in a short time.

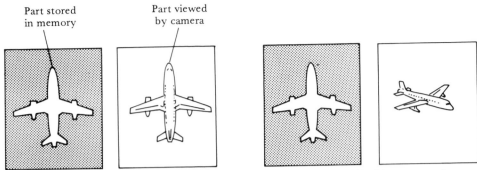

A. Image with same orientation   B. Image shifted by some angle

FIGURE 6-8   Template approach for object recognition (from Malcolm *Robotics*)

☐ SUMMARY

A number of methods or systems are used to accomplish the mating of computer and robot. The microprocessor is one device used to control robots.

Very intelligent robots have the ability to understand high-level languages. Almost every robot manufacturer has developed its own controller language. Some of the more frequently used languages are: VAL, HELP, AML, MCL, RPL, and RAIL. BASIC and CO-BOL are also used for some controllers.

A robot can be programmed by a computer. The program software (disk, magnetic tape, or other means) can be written and then adapted to the robot so that the robot does not have to be taken off the line during programming.

Interfacing is the means used to enable the robot to communicate with its controller and other parts of the work station. The microprocessor-controlled robot is able to communicate with other equipment around it by connections through ports.

ASCII code is the means by which the keyboard can be used to communicate with robot computers or microprocessors. Parallel ports are used when the computer and the machine it is controlling are separated by less than 50 feet. Serial ports are used when long-distance communication is necessary between units.

Information may be transmitted as changes in voltage or changes in current. The RS-232C standard and the TTL standard rely on voltage variations. The 60 mA and the 20 mA standards rely on current variations.

The controller has input ports for interfacing with various computer controls. RS-232C and RS-422 formats are used for this interfacing. A computer interface allows the programmer of the robotic system to program off-line.

Sensor information is converted to digital so the computer can handle it. Controllers have interfacing ports that provide for connecting sensors.

There are two types of program interfacing: service requests and robot requests. Each deals with interfacing with peripheral components and provides control during the period when the data are transmitted and received at the input ports.

Vision systems can provide the robot controller with information about the location, orientation, and type of part to be handled. Machine vision systems are used for recognition and verification of parts, for inspection and sorting of parts, for noncontact measurements, and for providing part position and orientation information to the robot controller. Edge detection and clustering are the two processes used for identification by the computer of the parts viewed by the vision camera. Much has to be done before vision is available for all robots at a reasonable price and with the reliability needed for daily production runs.

☐ KEY TERMS

ASCII   code used to transfer information from a keyboard to the processor, MPU, or CPU

CPU   central processing unit

interfacing   matching up of a device with a computer or microprocessor so they operate as one unit

language   the coded information fed to a computer to make it respond to commands and do certain operations

MPU   microprocessor unit

MVS   machine vision system; a method of giving the robot the ability to see

object recognition   ability of a robot to recognize certain shapes and to choose the right one for processing

off-line   refers to the programming of a robot by use of a computer and then placing the program in the robot's controller for action (The robot is not involved in the actual programming.)

parallel ports   method for connecting computers and peripheral devices so they can share data (uses eight or more wires)

RS-232C standard a standard that uses -3 to -25 volts for logic 1 and $+3$ to $+25$ volts for logic 0

serial ports method for connecting computers and peripheral devices so they can share data over distances of 50 feet or more (uses two wires)

60 mA standard a standard that uses 60 mA for logic 1 and zero for logic 0

software information programmed on floppy disks, hard disks, magnetic tape, or drums

TTL standard transistor–transistor logic standard that uses a 5 volt signal level for logic 1 and 0 for logic 0

20 mA standard a standard that uses 20 mA for logic 1 and zero for logic 0

☐ QUESTIONS

1. What is the purpose of a controller?

2. What is an MPU?

3. Where is the MPU located?

4. List six languages used by robots.

5. What is software?

6. Identify interfacing.

7. Where is the ASCII code used?

8. What is the difference between the TTL and RS-232C standards?

9. What is the difference between the 60 mA and 20 mA standards?

10. What is the strobe line?

11. What is the difference between the parallel port and the serial port?

12. How can you tell by looking if a port is parallel connected or serial connected?

13. Explain the difference between service requests and robot requests.

14. What is the main reason for putting vision on a robot?

15. How is the template method used for robot vision?

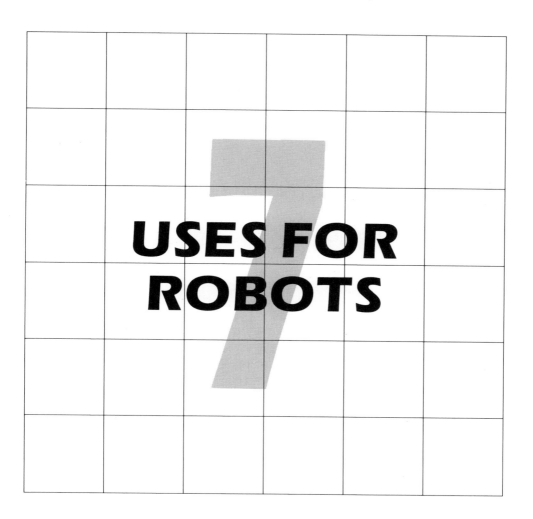

USES FOR ROBOTS

R obots are used for a variety of industrial purposes. Of course, we have to understand that *robot* means a device that can operate on its own without direct human supervision. Pick-and-place robots, the simplest type used in industry, can accomplish a number of jobs that humans do not especially enjoy doing. Humans tire easily when it comes to lifting and doing the same thing over and over. This can cause high absenteeism and affect the quality of the product being produced. Pick-and-place robots can go on all day loading and/or unloading without tiring.

Pick-and-place robots work well in handling materials. In most cases, they do not add value to the product by machining it or performing any other operation. They do the lifting, moving, stacking, and packing of parts or finished products.

More sophisticated machines are needed to do value-added work to a product or part. Value-added work is a task such as painting, welding, buffing, or assembling. More intelligent robots are employed to perform these tasks.

☐ LOADING AND UNLOADING

Work handling must be fast, accurate, smooth, and dependable if productivity improvements are to be achieved in modern industry. Even the simple picking up of a part and placing it on the line, as shown in Figure 7-1, must be efficient. Manufacturing efficiency increases in direct relationship to controlled work flow rates, good workpiece orientation, and precise part positioning. These essential factors allow plant processing equipment to be operated at full capacity for prolonged periods, with downtime limited to scheduled tool change and service-maintenance periods.

Lane Loader

A lane loader is another application for the pick-and-place robot. *Lane loaders* can be used to load and unload multiple part fixtures and balance flow rates between a fast machine and several slow machines or several slow to fast machines. See Figure 7-2.

Flow Line Transfer

Flow line transfer enables robots to be programmed to pick up two or more pieces at a time and transfer them off a machining line onto a sec-

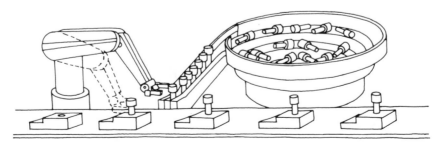

FIGURE 7-1   Pick-and-place robot used to perform simple, tedious materials-handling tasks (courtesy of FEEDBACK, INC.)

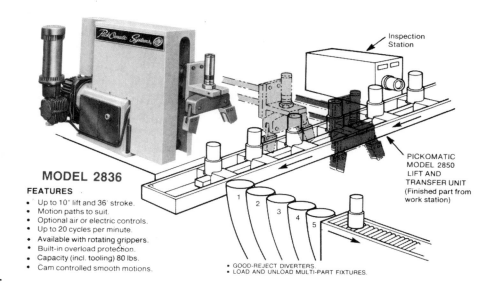

MODEL 2836

FEATURES

- Up to 10" lift and 36' stroke.
- Motion paths to suit.
- Optional air or electric controls.
- Up to 20 cycles per minute.
- Available with rotating grippers.
- Built-in overload protection.
- Capacity (incl. tooling) 80 lbs.
- Cam controlled smooth motions.

• GOOD-REJECT DIVERTERS.
• LOAD AND UNLOAD MULTI-PART FIXTURES.

FIGURE 7–2   Lane loader (courtesy of PICKOMATIC SYSTEMS®)

ond transfer line located parallel to the first one. Human effort in picking up and moving heavy loads is thereby eliminated. See Figure 7–3.

The main items to be concerned with here are the ability of the manipulator to move over the objects to be lifted and to swing over to put them down where they are needed. Another consideration is the weight of the package or part. The weight of the manipulator has to be factored into the total weight when you look at specifications for purchasing a robot to fit the job at hand.

Loading and unloading is the second largest use for robots. These machines do the jobs humans do not care to do since they may be in locations less than favorable for human habitation. The use of a robot to load and unload reduces the personnel injury rate and increases productivity. Excessive heat and noise, lack of light, and dirt and pollution do not bother the robot as it would humans. Using robots can lower production costs since safety equipment does not have to be purchased and maintained.

Machine Loading

One of the most important jobs the robot can do is machine loading. Most high-speed machines are tape numerically controlled (NC). The ability to make sure the machine is properly loaded is therefore of concern to any production planner. Numerically controlled machines work much faster than a human operator can and can improve the quality of

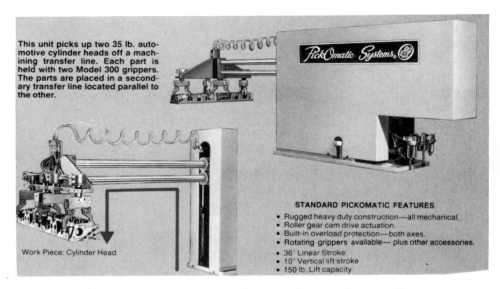

FIGURE 7-3   Robot used to transfer parts from one line to another (courtesy of PICKOMATIC SYSTEMS®)

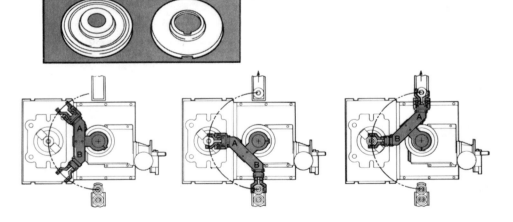

1. DWELL MODE
Work handler in dwell waits for ram to stroke and dies to form part.

2. UNLOAD MODE
Gripper "A" of handler removes formed part from dies while gripper "B" grips and removes blank from supply track.

3. LOAD MODE
Handler gripper "A" releases formed part to discharge track while gripper "B" loads blank into press dies. Handler returns to dwell mode position.

FIGURE 7-4   Automatic press loader (courtesy of PICKOMATIC SYSTEMS®)

the product, if properly fed with the parts they are machining. The cam-operated machine in Figure 7–4 shows how the parts handler can make it a smooth operation with no human involvement.

☐ MATERIALS HANDLING

Materials handling is the moving of materials about a manufacturing plant and can refer to raw materials as well as parts of a product that must be moved from one machine to another or from one part of the plant to another for further processing. Moving the material adds no value to the product, yet it has to be done. This means that industry must keep its materials-handling costs down.

Manufacturers depend heavily on conveyors to move parts and materials. Conveyors are synchronized with the robots so that parts or materials are where they are supposed to be when they are needed. The conveyors are connected to controllers so that their operations are synchronized with the machines doing the value-added work. See Figure 7–5.

Die Casting

The die casting industry was one of the first industries to put robots to work. *Die casting* involves working with very hot metal and presents a health hazard to humans. Tons of pressure are used to close two-piece molds, and the temperature of the metal being cast is such that it burns quickly and severely when it accidentally hits anything. Robots are ideal for doing the dirty work of die casting.

Two types of die casting machines in use are the hot chamber and the cold chamber. (The cold chamber is not really cold.) The hot chamber die casting machine melts its own metal and injects it into a two-piece mold under pressure. The material takes the shape of the mold, and the mold is cooled quickly. The part is removed and sent along for any further processing it needs. The hot chamber process normally uses zinc- and tin-based alloys, whereas with the cold chamber process, the metal is heated elsewhere and then is brought to the machine in a ladle. The cold chamber process usually casts bronze and brass at a higher temperature than the hot chamber process. See Figure 7–6.

The lost-wax process also is used in die casting. In this case, the pattern is made of wax. The wax is then coated with a material that will hold its shape when dried. The wax and its surrounding mold are heated until the wax melts, leaving a cavity in the mold the shape of the object that was made of wax. The cavity is filled with molten metal and spun around at high speed to cause the molten metal to cling to the inside of the mold. Once the metal has cooled sufficiently, the mold is broken and

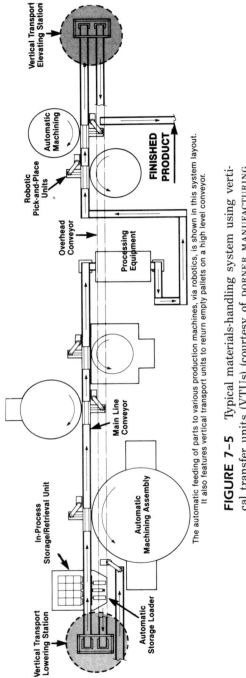

FIGURE 7–5 Typical materials-handling system using vertical transfer units (VTUs) (courtesy of DORNER MANUFACTURING CORP.)

The automatic feeding of parts to various production machines, via robotics, is shown in this system layout. It also features vertical transport units to return empty pallets on a high level conveyor.

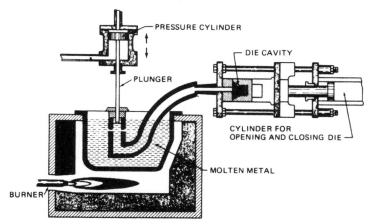

A. Hot chamber

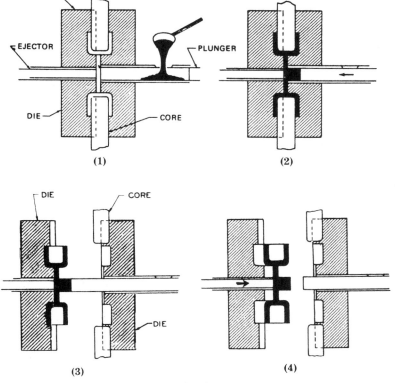

B. Cold chamber

FIGURE 7-6 Die casting processors (courtesy of NEW JERSEY ZINC CO.)

the metal part removed. This process is used for jewelry making and for very delicate work where fine details must be present in the finished product. See Figure 7–7.

Robots can do a number of operations that may be hazardous if done by humans. The robot can reach in and lubricate the mold when it opens, allowing the piece to fall out. It can take the piece and quench it or any of a dozen other operations that would endanger a human operator. The gases, dirt, dust, and pollution, as well as the heat and hot metal involved in die casting, all produce an undesirable atmosphere for humans. But the robot is ideally suited for the job. Figure 7–8 shows a machine feeding a machine, relieving humans of this repetitive, boring task.

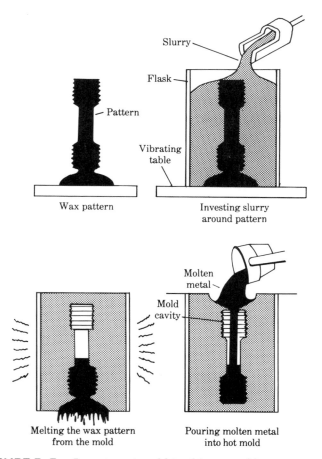

FIGURE 7–7   Investment mold (making a mold over a wax object)

FIGURE 7–8   Machine fed by machine (courtesy of I.S.I. MANU-FACTURING, INC.)

Palletizing

Stacking parts is easily automated, and robots were quickly used to do this type of work. *Palletizing*, or placing a series of boxes in a given pattern on a pallet, makes it easier for forklifts and other devices to pick up the pallet and move it to a shipping area or to another part of the manufacturing line. Picking up and placing boxes and parts into a specified arrangement can be very boring and tiring for humans. The robot, easily programmed to do such work all day long, never tires. A good example of a robot palletizing is shown in Figure 7-9, where the robot is placing bottles into a carton.

Just as parts can be taken from an assembly line and placed into a bin or box, they can be taken out of a box or bin and placed on the line. When they are removed from a box or bin and placed on the line for processing, the process is called *depalletizing*. See Figure 7-10.

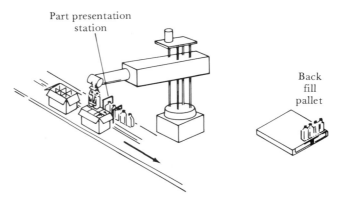

FIGURE 7-9 Robot used to place bottles in a carton for shipment (from Malcolm *Robotics*)

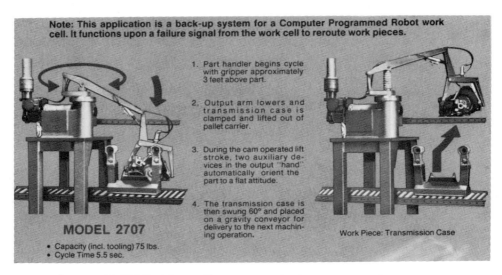

FIGURE 7-10 Pallet carrier unloader (courtesy of PICKOMATIC SYSTEMS®)

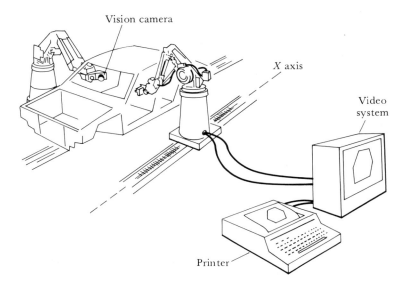

FIGURE 7–11 Line tracking robot with camera used to inspect car body weld joints (from Malcolm *Robotics*)

Line Tracking

In some instances, it is necessary for the robot to move with the production line. For instance, it may have to keep up with a part in order to inspect it. Stopping the line for such an inspection would increase production costs greatly. Thus, the robot must be placed on a parallel path with the moving part so it can do its job as the part moves to another location and operation. This is called *line tracking*. Figure 7–11 shows how a robot travels along with a car, inspecting the weld joints with machine vision.

Process Flow

In order for the production of any product to progress smoothly, you must have a regulated flow of parts and materials, and everything must be at the right place at the right time. This is referred to as *process flow.* A flowchart has to be made up so that the location of all parts is known at any given time. Then you have to arrange the timing so that all the parts arrive at a given location when needed. The demand for new parts must also be met by having raw materials available when needed. Keeping tools operational also has to be factored into the operation. All this is done by a flowchart for the entire plant operation. Then the flow of parts to a given area of supervision is drawn up for either robot or human supervision. Good planning makes for a good productivity record.

☐ FABRICATING

A robot can do a number of the required operations in *fabricating* parts or a product. Some of the more prevalent manufacturing processes being performed by robots are routing, milling, drilling, grinding, polishing, deburring, sanding, and riveting. See Figure 7–12, which shows a complex double gripping process in which a hole is drilled and then deburred perpendicular to the main axis of the part. The robot also can change tooling, can change fixtures, and has an increased accuracy and repeatability.

Improved sensory and adaptive control causes a robot to be used in different manufacturing processes. Robots have increased productiv-

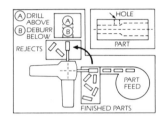

A. Overall layout

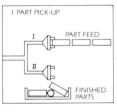
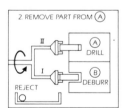
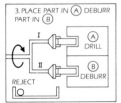
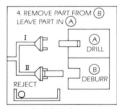
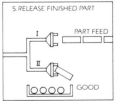

1. A part is picked from the part feeder. The robot then base rotates 90° while wrist rotating gripper II to the top.

2. The robot exends, and gripper II grasps the part from drill unit A. The robot arm retracts and rotates gripper I to the top. At this point gripper I holds the raw part and gripper II holds the drilled part.

3. The robot arm extends. The jaws of drill A grasp the raw part while B grasps, deburrs, and releases the drilled part in gripper II.

4. The robot arm retracts leaving the part in A and removes the finished part from deburring unit B.

5. After a 90° base rotation, the finished part is released from gripper II. The cycle then repeats. (*Note:* Parts in the wrong orientation from the feeder are released just prior to step 3.)

B. Side views of the main steps

FIGURE 7–12   A double gripping application used to speed up the manufacturing process (courtesy of SCHRADER-BELLOWS)

ity and improved the quality of the parts produced. In addition, the robot takes the worker out of the dust, noise, and polluted air.

☐ ASSEMBLING

Robots have been used in *assembling* calculators, watches, printed circuit boards, electric motors, and alternators for automobiles. The future will see robots assembling anything that can be broken down into simple step-by-step operations, and that includes almost everything.

More and more attention is being paid to the possibilities of using robots for the tedious jobs of assembling small and/or large parts. See Figure 7–13 for an illustration of how a robot is used to fabricate printed circuit boards.

One of the problems with converting a product to automated assembly is its design. Some products are not designed to be assembled by machines. See Figure 7–14. Designers must take a close look at each product and be aware of the ability of the machines that do the assembly work. By studying the design requirements of a product and the ability of the robot or other machines, it is possible to design a better-

FIGURE 7–13   Robot used to fabricate printed circuit boards
(courtesy of FARED ROBOT SYSTEMS, INC.)

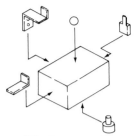
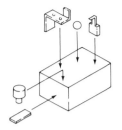

A. Difficult to assemble
   because of part orientation

B. Preferred assembly
   because of new orientation

C. Difficult to grip
   because no lip provided

D. Preferred assembly
   because lip provided

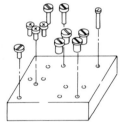
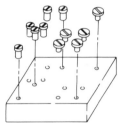

E. Difficult to automate
   because of too many
   screw variations

F. Preferred assembly
   because of fewer
   screw variations

FIGURE 7–14   Product design – an important factor in automated assembly work (from Malcolm *Robotics*)

quality device and have it assembled by machine. The ability to assemble an electric motor without a human hand touching it is worthy of note. The ability of machines to make alternators for automobiles is another application that keeps quality and productivity up and cost down. Robot assembly relieves workers from performing boring, repetitive tasks day after day. The quality of the product can be improved by having the robot work at a constant pace. The quality of the product improves because the tolerances are closer when a machine is used to handle the parts.

☐ PAINTING

The robot's wrist is the usual location for a spray painting gun. A human operator takes the robot arm through the motions of spray painting. A continuous-path robot with lots of memory is usually used. The robot memorizes the paths and then repeats them. It has to be debugged after a few sweeps since both the correct and incorrect moves of the human operator were memorized and stored. The program has to be checked often before putting the robot to work. See Figure 7–15.

Spray painting by robot has several advantages. One of them is the removal of the human operator from the fumes, making it possible to use different types of paints. Many of today's coatings are toxic and need to be applied by the spray arm of a robot since humans would be injured if caught inside the enclosure with the fumes. Another advantage is that robots can spray paint very thin lines with accuracy. A robot can paint pin-stripping on automobiles all day without faltering. Areas that could not be reached by a human painter can be reached by high-technology, articulate robots because they have six (some up to nine) axes.

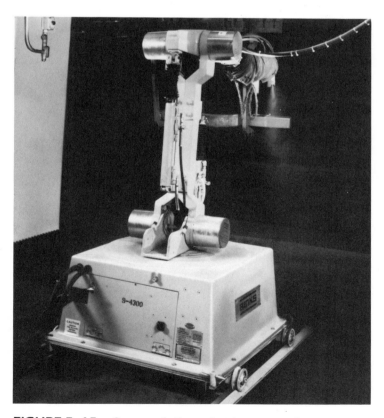

FIGURE 7–15 Spray painting robot (courtesy of BINKS MANU-FACTURING COMPANY)

☐ WELDING

One of the largest uses for robots is welding. Industry uses robots for spot, stud, stick, metal-inert gas (*MIG*), and tungsten-inert gas (*TIG*) welding. (TIG is a form of stick welding.) Spot and stud welding are resistance welding processes. Stick, MIG, and TIG are arc welding processes.

Spot welding is the most common use of welding because it is the easiest and is in the greatest demand. When the assembly operation begins on automobiles, spot welding is used to hold the pieces of sheet metal together. Hundreds of spot welds are made to keep the pieces of metal together and to provide a tight body without rattles and squeaks. Spot welding can be performed as the automobile body moves along, or the welder-robot can be stationary.

The most common type of arc welding is MIG. The welding station consists of a wire feeder, an inert gas supply, a gasflow meter, and a welding gun. The welding is a continuous process once it starts. An inert gas is brought to the spot where the welding is taking place to shield the metal from oxygen in the air while it is very hot and in a molten state. The current or amperage is controlled by how fast the wire is fed to the joint. See Figure 7–16. Remote welding by robot has been per-

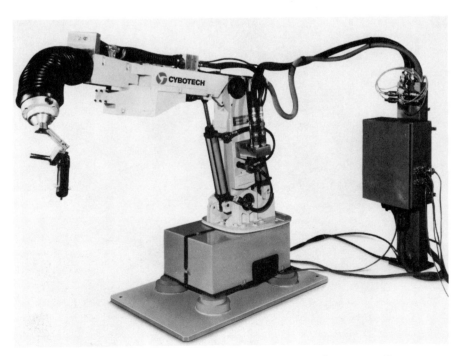

FIGURE 7–16   Arc welding robot (courtesy of CYBOTECH® INDUSTRIAL ROBOTS)

fected to such an extent that the wire feed speed can be controlled within ± 1 inch per minute and the welding voltage within ± 0.1 volt.

One of the biggest problems with machine welding is the butting together of the two materials to be joined. The groove must be accurate and fit smoothly. The robot does not, in most instances, have the ability to adjust the welding path or wire feed or to change the current according to demands as a human operator would.

☐ INSPECTING AND TESTING

Other tasks a robot can perform without tiring are testing and inspecting. Humans have a tendency to tire after doing a repetitive job for some time. Their efficiency and objectivity are both compromised. Thus, a robot with the ability to test products without tiring would, in most instances, produce a better-quality product. Testing and inspecting can be done with a number of robots. Robots can pick up a part and place it into an inspection gage or a more complicated device for a go/no–go testing situation. Robots also may have vision that will allow them to test up to one thousand points on a part in one minute.

Inspection is less complicated than testing. It involves taking a look at a part, gaging it, and then discarding it if it does not fit the established limits. Robots have a number of advantages when it comes to inspections. They can perform the inspection as the part moves along, do the same thing over and over without tiring, be used to test 100 percent of the product line, and obtain consistent results. See Figure 7–17.

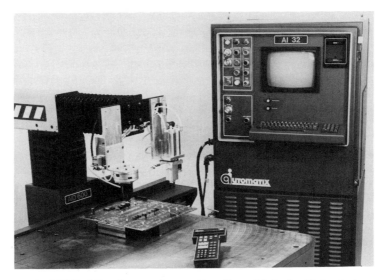

FIGURE 7–17   Inspection station (from Malcolm *Robotics*)

☐ FUTURE OF FLEXIBLE AUTOMATION

The computer-integrated manufacturing (*CIM*) factory of the future will look quite different from the factory of today. It will be based on the integration of the traditional or process-based technology of today, with the emerging software- or systems-based technology of today and tomorrow. See Figure 7–18 for an example of the factory of the future.

Objectives of CIM

Seven objectives in setting up a computer-integrated manufacturing process or method of organization for making a product are as follows.

1. *Obtain an economic order quantity approaching one.* While it probably will be some time before an economic order quantity of one is feasible for manufacturers, it should be their ultimate goal. It is the key to controlling the cost of inventories of finished goods and work in process and will allow total flexibility in the factory.
2. *Approach a set-up time of zero.* The development of universal, multipurpose fixtures, tool changing systems with unlimited numbers of tools, storage of part-machining and assembly programs in computer memory, set-up probes, and other innovations are drastically reducing set-up time.
3. *Obtain family-of-parts programming and production.* The rationale for manufacturing processes is to start with grouping parts with similar characteristics into families. Parts related by design parameters and/or manufacturing characteristics are grouped together for optimized manufacturing and/or assembly in work cells.
4. *Integrate design and manufacturing.* A CAD-CAM system is needed to make sure the computer-aided design (CAD) and computer-assisted manufacturing (CAM) package is such that it is an integrated, interactive system in which the parts designs are optimized to be manufactured efficiently, and the manufacturing process has the flexibility to accommodate relatively smaller numbers of a larger variety of parts.
5. *Establish inventory integrity and just-in-time parts delivery.* Materials procurement and handling are, typically, the largest manufacturing cost items (50 percent versus 10 percent for direct labor and 15 percent for indirect labor). Just-in-time (JIT) production minimizes handling, and work-in-process (WIP) inventory can dramatically reduce capital requirements, improve cash flow, and reduce the break-even point.
6. *Establish absolute control of the total process.* In a properly de-

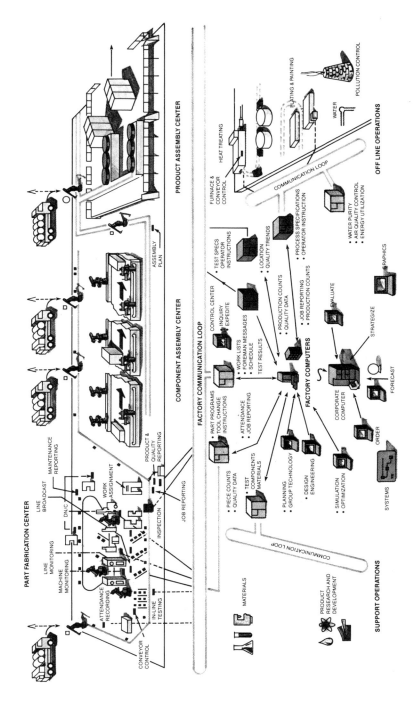

FIGURE 7-18 Factory of the future (courtesy of GCA CORPORA-TION/INDUSTRIAL SYSTEMS GROUP)

signed CIM system, work cells are integrated with other shop floor and factory management computer systems, as well as the central database management system. Interfaces between all subsystems communicate with proven, portable software. All gaps between islands of automation are bridged so that the end-user can replan, reschedule, and reprogram in real time. In a properly designed CIM system, the data communication loop is completely closed. Management has complete control of the process at the center level.

7. *Maximize efficient use of available workspace.* Robotized work stations in flexible manufacturing and assembly systems utilize minimum floor space. Pallet shuttles and vertical, gravity-feed parts delivery systems also make efficient use of space.

These objectives are listed here courtesy of GCA Corporation/Industrial Systems Group, Naperville, Illinois.

☐ FUTURE OF ROBOTICS

Manufacturers in the United States have been more successful in building robots than in selling them. This appears to be the sum of things at the moment. Once we get geared up for production, the possibility of overproduction always exists. There is some resistance to the adoption of robots for factory work and other tasks. Some people are very suspicious when it comes to being replaced by machines; therefore, every type of roadblock may be thrown up to impede the progress of the new approach to manufacturing. However, there are some positive aspects to be gained from robotization.

Vision systems are too expensive and slow at the moment but will improve as research and development progress to meet the demand for better-quality recognition by robots. Tactile sensing is in need of development and will continue to develop as manufacturers move to make sure their product is the best. In the future, robots will become smaller, more accurate, and faster and will have better repeatability. Better and easier programming can be expected for robots of the future.

Research projects are under way at a number of colleges and universities — the University of Michigan, Stanford, and the University of Rhode Island, to name but a few. Others are also involved in improving the robot, its vision, and its contribution to the social health of the nation.

☐ SOCIAL IMPACT OF ROBOTS

Union and industry spokespersons report that the march of industrial robots onto assembly lines will cost hundreds of thousands of jobs in the next few years but may save some companies from going under. A study by the United Auto Workers' Union estimated that as many as 225,800 jobs could be lost in the auto industry by 1995. Throughout the auto industry, between 48,300 and 58,800 robots will be used in factories by 1995, compared with 9,200 to 9,800 in 1985. The robot population of the United States in 1995 will total more than 128,000 with each machine on average replacing two workers.

☐ NEW USES AND NEW FORMS

The U.S. Army is developing a six-legged robot for walking over various types of terrain and going places that tanks cannot venture. Some of the research for this project will aid in making a walking robot with legs instead of wheels. Balance problems and movement of the legs will be solved with more applied research. Better vision and tactile sensors will also be developed for this project.

Brain surgery is already being aided by robots in the operating room. When lasers and robots, along with CAT scans, team up, it will mean a new approach to solving some surgical problems. The future holds much promise for robots and for those who work with them or on them.

☐ SUMMARY

Work handling must be fast, accurate, smooth, and dependable if productivity improvements are to be achieved. Pick-and-place robots efficiently accomplish many loading and unloading jobs that are slow, tedious, and boring for humans.

A lane loader is another application for the pick-and-place robot. Lane loaders can be used to load and unload multiple part fixtures and balance flow rates between fast and slow machines.

Flow line transfer is also important in keeping the line moving and production going. Twin-armed or even three-armed robots can handle this situation well. Loading and unloading is one of the largest uses for robots.

Machine loading is one of the important jobs for a robot. It prevents humans from getting fingers caught in a press or forge. Materials handling, that is, the moving of materials about a manufacturing plant, is one of the expense items in the manufacture of any product. If the cost of handling materials can be reduced, so can the price of the product.

Die casting was one of the first industries to use robots. Robots can handle the hot materials without having to react to the heat, dirt, pollution, and lack of lighting. Robots can do things in the casting process that humans are unable to do.

The stacking of parts is easily automated, and the robot does the job well. Parts can be taken from an assembly line and placed in a bin or box (palletizing), or they can be taken out of a box or off a pallet by a robot (depalletizing).

Line tracking is done by the robot moving along with the line and performing its job as it moves along. Process flow is the moving of parts and materials in an orderly and timely manner.

Fabricating can also be done by robots. Everything from drilling, riveting, sanding, deburring, and grinding to polishing can be done by robots. They can also assemble products partially or totally without human supervision.

Painting and welding are two of the value-added processes that robots can do. They are easily adapted to welding and spray painting. Inspecting and testing are also easily automated, with robots designed to do the job without tiring. Robots have a tendency to improve the quality of the finished product since they are capable of 100 percent inspections.

The computer-integrated manufacturing (CIM) plant of the future will look quite different from the factory of today. It will be based on the integration of the traditional or process-based technology of today, with the emerging software- or systems-based technology of today and tomorrow. The seven objectives of CIM provide for cost control, innovative and integrated procedures, and complete and efficient design and control systems.

The future holds promise for six-legged robots and for robots that aid in brain surgery. The movement toward robotization has only begun. Your imagination is the only limit.

□ KEY TERMS

assembling   putting together

CIM   computer-integrated manufacturing; one of a number of proposed organizational methods for manufacturing products in large quantities

depalletizing   the use of a robot to unload parts or boxes off a pallet

die casting   the use of dies to form hot metal into desired shapes

fabricating   making something

flow line transfer   the use of a robot to pick up two or more pieces at a time and transfer them off a machining line onto a second transfer line located parallel to the first one

lane loaders   pick-and-place robots used to adjust the feed between fast and slow or slow and fast lines

line tracking   the process of having a robot move along with the production line and do its work as it moves with the line

MIG   type of metal-in-gas welding where an inert gas surrounds the welding spot or area while it is in the molten state

palletizing   the use of a robot to stack parts or boxes on a pallet

process flow   the orderly flow of parts and materials to keep production going

spot welding   welding a small spot between two electrodes of a spot welder; melting metal with a surge of high current through the metal

TIG   type of stick welding; tungsten-in-gas

☐ QUESTIONS

1. What are the essential factors for work handling?

2. What is a lane loader?

3. What is flow line transfer?

4. What does NC stand for?

5. What is a conveyor?

6. What is meant by the term *value-added?*

7. How is die casting done?

8. What are two types of die casting?

9. What is palletizing?

10. Why are robots so well adapted to welding and spray painting?

11. List three types of welding that robots can do.

12. What are the objectives of CIM?

MANUFACTURERS' EQUIPMENT

Now that you have read this book and acquired some knowledge of the fundamentals of robots and robotics, the next questions to consider are what do other systems look like, and what can they do? To find the answers, we usually look at manufacturers' catalogs and check out the characteristics, capabilities, and technical qualities of robotic equipment manufactured in the United States, Europe, and Asia.

Robots are made by many companies the world over. A great many of them are made for the manufacture of automobiles. They do welding, painting, and assembly work very well in Italy, Sweden, Germany, Japan, Korea, and other countries. Robots are also used in the assembly of printed circuit boards for all types of electronic equipment. The electronics industry is one that can brag about lowering the price of its products rather than increasing them every year. In fact, you expect electronic equipment to decrease in price with time. One of the reasons for this is the elimination of costly hand labor. The robot has become the workhorse of the electronics industry as well as the auto industry. You will find many of the robots shown here heavily involved and utilized in these two areas of manufacturing.

No one name brand of robots has captured the market. There are a number of manufacturers in the business. However, the number decreases all the time. Robot manufacturers are merging, being bought out, or declaring bankruptcy in great numbers. The shake-out in this industry is similar to that in the computer industry. Some of the more well-established firms and their literature has been selected for discussion here.

This chapter will acquaint you with a broad cross-section of what is available. Computers and microprocessors are mentioned, but you should consult other sources for more detailed information on them.

□ SELECTED MANUFACTURERS

A list of selected U.S. manufacturers and their location follows.

- Binks Manufacturing Company (Franklin Park, Illinois)
- Cincinnati Milacron, Inc. (Lebanon, Ohio)
- Comau Productivity Systems, Inc. (Troy, Michigan)
- Cybotech Corporation (Indianapolis, Indiana)
- Elicon (Brea, California)
- ESAB North America, Inc. (Fort Collins, Colorado)
- Fared Robot Systems, Inc. (Denver, Colorado)
- Feedback, Inc. (Berkeley Heights, New Jersey)
- GCA Corporation/Industrial Systems Group (Naperville, Illinois)
- Hobart Brothers Company (Troy, Ohio)
- International Business Machines Corporation/Manufacturing Systems Products (Boca Raton, Florida)
- International Robomation/Intelligence (Carlsbad, California)
- I.S.I. Manufacturing, Inc. (Fraser, Michigan)
- Lamson Corporation (Syracuse, New York)
- Mack Corporation (Flagstaff, Arizona)
- Microbot, Inc. (Mountain View, California)
- Pick-O-matic Systems, Inc. (Sterling Heights, Michigan)
- Prab Robots, Inc. (Kalamazoo, Michigan)

- Schrader-Bellows, Division of Parker-Hannifin (Akron, Ohio)
- Seiko Instruments USA, Inc. (Torrance, California)
- Transcom, Inc. (Mentor, Ohio)
- Thermwood Robotics (Dale, Indiana)
- Unimation Incorporated, a Westinghouse Company (Danbury, Connecticut)
- Yaskawa Electric America, Inc. (Northbrook, Illinois)

☐ SELECTED EQUIPMENT

Illustrative material provided by manufacturers is presented in the following sections. These manufacturers were very enthusiastic and willing to supply data. We have tried to represent them properly, with all the information being taken from their brochures and other available instruction sheets. You will find specification sheets for some of the products mentioned here at the end of this chapter.

Binks Manufacturing Company
9201 West Belmont Avenue
Franklin Park, Illinois   60131

88-800 ROBOT

The Binks 88-800 is the only robot designed and manufactured in the United States specifically for spray painting. It has received excellent acceptance. It is noted for its flexibility, ease of operation, and reasonable price. (See Figure 1-8.)

The advanced articulation of the 88-800 lets it apply uniform coatings to virtually any product that can be sprayed manually. Its 18-pound capacity permits the use of a wide range of spray equipment, including air, airless, electrostatic, and plural component spray guns.

A flexible, corrugated plastic shroud protects moving parts from dust, dirt, and overspray — a major advantage in the application of frits in porcelain enameling. The clean, unobstructed lines of the arm allow it to reach into recessed or enclosed areas to paint interior surfaces. Factory Mutual Approval has been obtained for use of the 88-800 in hazardous locations.

The Binks 88-800 is a servo-controlled, 6-axes (additional axes optional), continuous-path robot with features not previously available in spray painting robots. Its solid-state computer control system can store up to eight programs. An optional disk memory can store hundreds of programs and forty hours of total programming time.

A unique editing system makes it possible to correct a program, by either gun motion or triggering, without removing the robot from production. An exclusive adaptive correction system lets the machine perform any program sequence while automatically adjusting for changes in temperature, fluid viscosity, load, and other factors affecting its operation.

The 88-800 is easily programmed by manually leading it through the spraying operations to be performed. The lightweight, counterbalanced arm allows free movement of the spray gun during programming sequences.

The completely solid-state control has no moving parts and is tolerant of humidity, dust, and temperature. At the push of a button, a diagnostic system quickly isolates electronic problems. The mechanical design of the 88-800 provides major economies. Its unique editing control permits the system to be fine-tuned to minimize material usage and rejects.

Manipulator Unit: The manipulator unit is a hydraulically driven, servo-controlled, six-axes unit that will accurately duplicate the motions of a skilled painter. See Figure 8-1. The entire spraying sequence can be recorded by continually depressing the program trigger. A point-to-point program may be recorded by turning the switch on and off at each of the desired points. The spray gun trigger switch is located on the front handle. The manipulator can be used in Class I, II, III, Division I, Groups D, F, and G hazardous locations. It is Factory Mutual Approved.

Robot Controller: The solid-state control system eliminates the normal problems associated with moving tape readers, card readers, magnetic tape cartridges, and reels. The control system includes a complete electronic diagnostic system. The system will operate in temperatures between 40°F and 120°F.

Hydraulic Power Supply: The hydraulic power supply is a standard-type unit complete with filter, pressure regulator, and water/oil heat exchanger. The normal operating range is 600–700 psi. Built-in fault detection is included and is interfaced with the robot controller to monitor oil, filter, and temperature.

Safety Fencing: The safety fence will prevent personnel from entering the area while the robot is in operation. If the gate to the robot area is opened, hydraulic power to the robot is shut down. The fence is constructed of seven-foot high wire mesh, with gate and limit switches. See Figure 8-2 for a typical installation.

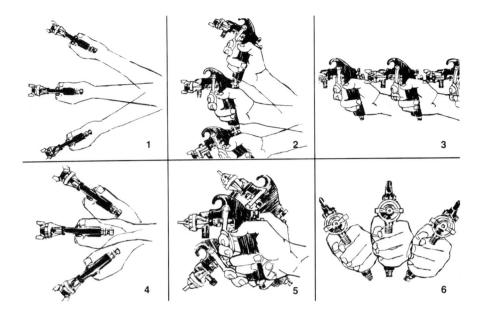

1. **Rotary Stroke 135°**
2. **Vertical Stroke 84″**
3. **Horizontal Stroke 48″**

4. **Horizontal Wrist Movement 180°**
5. **Vertical Wrist Movement 180°**
6. **Wrist Roll 270°**

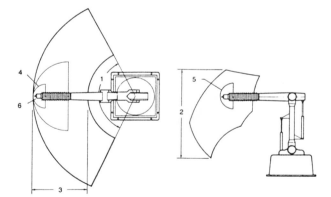

FIGURE 8-1   The Binks robot 88-800 showing the six actions available in hand spray gun operations (courtesy of BINKS MANUFACTURING COMPANY)

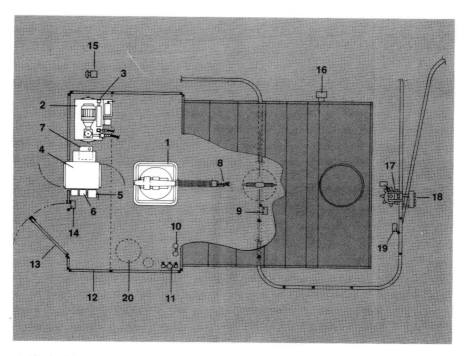

1. Manipulator
2. Hydraulic Power Supply
3. Water/Oil Heat Exchanger
4. Control Enclosure, Floor Standing
   Includes: a) Robot Controller, b) Hard Disk Memory,
   c) CRT Terminal, d) Air Conditioner
   for Enclosure.
5. System Disconnect
6. Transformers 220/110 or 440/110
7. Electrostatic Power Supply
8. Spray Gun
9. Program Start Limit Switch
10. Gun Trigger Solenoid Valves

11. Compressed Air Regulators
    and Filter Assembly
12. Wire Mesh Enclosure
13. Enclosure Gate
14. Gate Interlock Limit Switch
15. Remote Emergency Stop P.B. Station
16. Exhaust Flow Switch
17. Memory Parasitic Drive Assembly
18. Program Identification Manual Input
    System
19. Program "Enter" Limit Switch
20. Material Supply

FIGURE 8–2 Typical 88-800 robot installation (courtesy of
BINKS MANUFACTURING COMPANY)

Cincinnati Milacron, Inc.
Industrial Robot Division
215 South West Street
Lebanon, Ohio   45053

T³363 and T³746 ROBOTS

A "simple robot for simple tasks" is the way the T³363 robot is described. It was designed and built to bring the productivity of off-the-shelf robotic automation to simple tasks, such as machine tending and medium-duty materials handling. This all-electric robot operates on three axes of movement — two linear and one rotary. For even more flexibility, an optional fourth servo axis is available for pitch or yaw. (See Figure 2–9.) The compact design of the robot puts its total working volume within 18 inches of the floor and allows it to deliver a full 300° of rotation, with a payload of up to 110 pounds.

Cincinnati Milacron makes a number of robots for a wide variety of jobs. The T³746 is a typical type. Its work envelope is shown in Figure 8–3. Spec Sheet #1 gives all the details of its capabilities. It is an

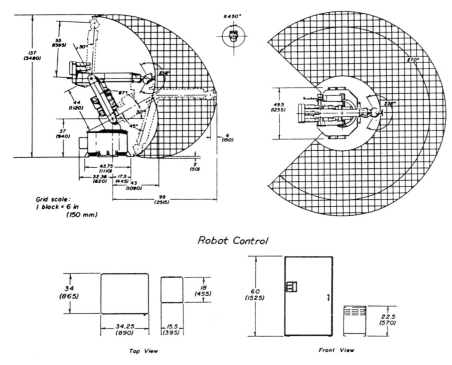

FIGURE 8–3   Work envelope and control cabinet for T³746 robot (courtesy of CINCINNATI MILACRON)

electric-driven, computer-controlled, versatile industrial robot. Its DC motor drives and its 70-pound load capacity enable it to boost productivity in a wide variety of process applications such as:

arc welding	routing
inspection	deburring
materials handling	grinding
polishing	drilling
sealant application	

Control: Cincinnati Milacron's microprocessor-based ACRA-MATIC Version 4 control and operator-friendly software make programming easy. There is no need for computer experience — just a knowledge of the job the robot is to perform.

Simple use of the lightweight, hand-held teach pendant leads the robot through its required moves. See Figure 8–4. The controlled path motion feature automatically coordinates all six axes to move the robot from one point to the next at the velocity selected in world coordinates. An average of 3,000 points can be programmed and stored within the control memory.

The T^{3}746 can be used as the central element in a robotic arc welding system. With this system, the power supply, wire feeder, positioning tables, and associated hardware are all interfaced with the robot through its own computer control.

FIGURE 8–4   Teach pendant used with the T^{3}746 robot (courtesy of CINCINNATI MILACRON)

Comau, S.p.A.
Strada Borgaretto
22
Torino, Italy

Comau Productivity
Systems, Inc.
750 Stephenson
Highway
Troy, Michigan   48083

SMART™ ROBOT

SMART is the six-axes, all-electric industrial robot designed by Comau of Torino, Italy. It is mounted on the floor, upside down, sideways, or in any position to best meet the application requirements. This robot is used extensively in the manufacture of automobiles.

The operator can program and operate the robot from either the operator's panel and/or a lightweight, hand-held microcomputer-based pendant unit (programming terminal).

Robot programming is performed using the versatile Comau PDL (process description language), a high-level language that permits the user to define motion sequences and logic sequences; activate, deactivate, and synchronize independent processes; and easily define complex cycles (e.g., palletizing or tasks in which the decision about the actual path of the arm is made at run time).

Robot motion control, both during programming and automatic operation, is accomplished by point interpolation using one of the following methods:

> . . . Points are generated for simultaneous, coordinated motion of all axes traversing the tool center point (TCP) through the shortest path between programmed positions (point-to-point motion in the robot coordinate system).
>
> . . . Points are generated for coordinated motion of all axes, moving the TCP in a straight line from one programmed position to the next, while performing programmed tool orientation (controlled path motion in the Cartesian world coordinate system X_B, Y_B, Z_B, originating in the robot base center). See Figure 8–5.
>
> . . . Points are generated for coordinated motion of all axes, moving the TCP in a straight line from one programmed position to the next, always maintaining the taught tool orientation (controlled path motion in the Cartesian tool coordinate system X_T, Y_T, Z_T, originating in the TCP itself). See Spec Sheet #2 for robot characteristics.

In controlled path motion, the ultimate servo commands are derived from point interpolation in the real world, through the transforma-

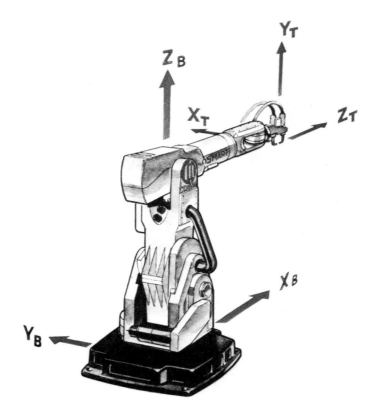

FIGURE 8–5 Six axes of the SMART robot (courtesy of COMAU PRODUCTIVITY SYSTEMS, INC.)

tion from fixed world coordinates to robot coordinates. Advanced algorithms provide a higher-level feedback on TCP velocity and position, thus allowing full-speed positioning (or positioning at the selected speed), according to servo torque limits.

Controller: The computerized control system is housed in a single cabinet with cables that run 5 meters to 40 meters. See Figure 8–6.

The operator's panel has a 9-inch CRT, alphanumeric keyboard, and control panel. Optionally, a serial printer can be attached for printing of source programs and/or machine parameters.

All robot axes are driven by DC servo motors. See Figure 8–7, which shows the work envelope for the SMART robot. Resolvers are mounted on the motor shaft and provide an absolute cyclic position feedback. The axes zeroing procedure is necessary only in case of emergency manual operations. Compact, modular, multiaxis PWM transistor DC motor drives are mounted in the controller cabinet.

FIGURE 8-6   Control console for the SMART robot (courtesy of COMAU PRODUCTIVITY SYSTEMS, INC.)

Applications: Connected to a vision system, the SMART robot can pick semisorted parts and feed conveyors, intermediate buffers, transportation systems, and so forth. In such cases, the robot's job is to boost the productivity of an entire system of machines.

The SMART work envelope is well suited to spot weld a wide range of subassemblies, such as body sides, doors, hoods, floors, and complete frames. See Figures 8-8 to 8-11.

The COMAU **SMART** Robot has an articulated anthropomorphic configuration.

The currently available versions are.
- **SMART 6.50R** Rotating Base (shown in the Figure).
- **SMART 6.50T** Traversing Base.
- **SMART 5.50R** (without the 4-Axis).
- **SMART 5.50T** (without the 4-Axis).

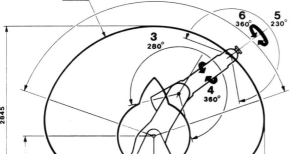

Each axis is driven by a DC servo motor through a gear reduction mechanism; the motor, tachometer, resolver, and brake are attached to a common shaft.

The main axes are pneumatically or weight counterbalanced. Wiring and air supply are internal to the robot arm.

Each axis has a microswitch that establishes the zero position. Axes zeroing is required only after an emergency stop or after a manual operation (with the controller electronics turned off). The 1, 2, 3 and 5 axes have overtravel mechanical stops; on 1 and 3-Axes they can be adjusted to limit axes travel.

Positioning repeatability is ± 0,4 mm.

Base floor space is 1150 × 1150 mm.

The horizontal reach to tool flange is 1917 mm, the vertical reach is 3000 mm.

The load capacity is 50 kg (400 mm from wrist center and 150 mm offset), or 60 kg (330 mm).

Axes motion range and speed are:

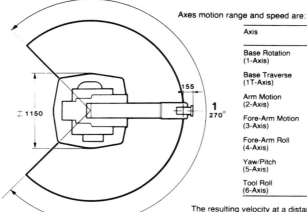

Axis	Motion Range	Speed
Base Rotation (1-Axis)	270°	78°/sec
Base Traverse (1T-Axis)	2.5 m (or multiples)	1 m/sec
Arm Motion (2-Axis)	140°	80°/sec
Fore-Arm Motion (3-Axis)	280°	102°/sec
Fore-Arm Roll (4-Axis)	360°	138°/sec
Yaw/Pitch (5-Axis)	230°	136°/sec
Tool Roll (6-Axis)	360°	144°/sec

The resulting velocity at a distance of 1.9 m from the base center is approximately 2.6 m/sec.

FIGURE 8-7   Work envelope for the SMART robot (courtesy of COMAU PRODUCTIVITY SYSTEMS, INC.)

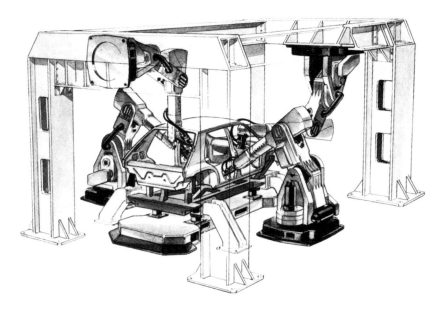

FIGURE 8-8   Using the SMART robot in various mounts
(courtesy of COMAU PRODUCTIVITY SYSTEMS, INC.)

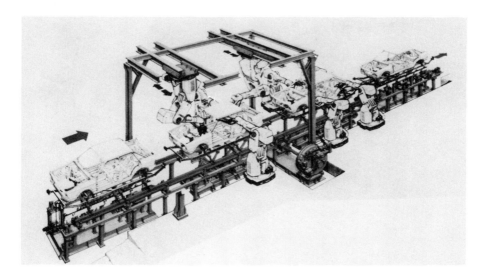

A. Using eight SMART robots to weld four different types of body shells

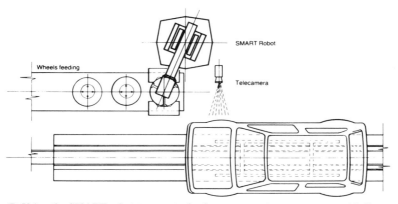

B. Using the SMART robot to mount wheels on a car as it moves on assembly line

FIGURE 8-9   Assembly line applications of the SMART robot
(courtesy of COMAU PRODUCTIVITY SYSTEMS, INC.)

A. Robot picking up semi-ordered parts and placing them on a conveyor

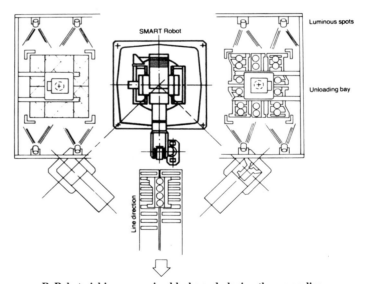

B. Robot picking up engine blocks and placing them on a line

FIGURE 8–10 Parts handling applications of the SMART robot (courtesy of COMAU PRODUCTIVITY SYSTEMS, INC.)

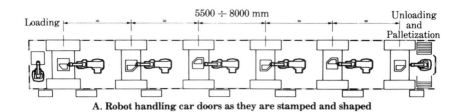

A. Robot handling car doors as they are stamped and shaped

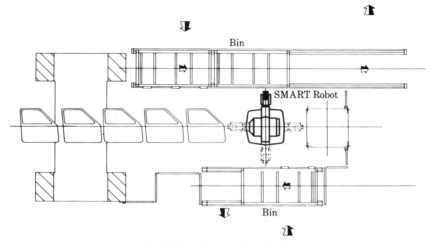

B. Robot putting car doors in bins

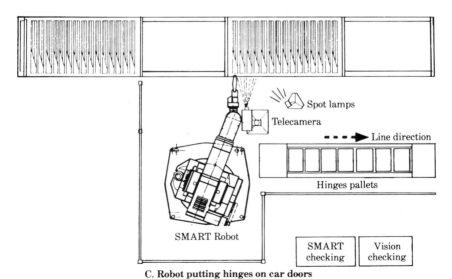

C. Robot putting hinges on car doors

FIGURE 8-11   Automotive application of the SMART robot
(courtesy of COMAU PRODUCTIVITY SYSTEMS, INC.)

Cybotech Corporation
P.O. Box 88514
Indianapolis, Indiana   46208

H80, G80, V80, V15, and P15
ROBOTS

Cybotech robot systems are capable of handling painting, spot and arc welding, assembly, routing, drilling, and numerous other applications.

The H80 has five rotary axes and one translational axis. Either hydraulic or electric, it is ideal for applications requiring work along horizontal surfaces or in situations involving vertical obstacles.

The G80 gantry has three rotary axes and three translational axes. See Figures 8–12 and 8–13. The size of its work envelope depends on the dimensions of the robot and is virtually unlimited. Either electric or hydraulic, the G80 is well suited to applications such as transfer line manufacturing; flat, downhand, open-arc welding; and other situations involving horizontally oriented processes.

The G80 is a versatile, precise industrial robot built for applications with tool and part loads of up to 175 pounds. The gantry configuration is ideally suited for applications where overhead mounting is advantageous, such as in transfer line operation. The G80 is available in either electric or hydraulic versions. It can be equipped to perform tasks such as arc welding, spot welding, drilling, and cutting with a plasma arc or water jet torch. The G80 design affords maximum productivity with an 80 kg load. Repeatability is better than ± 0.2 mm (0.008 inch).

Multiple G80 robots can be suspended from the same gantry framework. This is appropriate for assembly line applications where

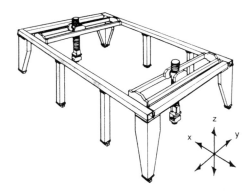

FIGURE 8–12   The G80 robot in double gantry configuration (courtesy of CYBOTECH INDUSTRIAL ROBOTS)

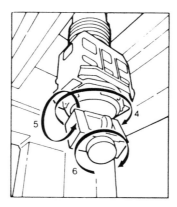

FIGURE 8-13   The G80 robot's three rotary axes (courtesy of
CYBOTECH INDUSTRIAL ROBOTS)

similar or identical operations are required on both sides of a workpiece.
Spec Sheet #3 shows the applications possibilities for this type of robot.

The V80 is a hydraulic robot with six rotational axes. It is well
adapted to applications where work must be reached from the top or
where obstacles must be avoided. (See Figure 3-15.)

The V15 has either five or six rotational axes. This compact, accurate electric robot is suited to numerous manufacturing operations
such as welding, grinding, drilling, cutting, and parts handling.

The P15 is a hydraulic robot with seven rotational axes. It is ideal
for painting and coating because it is large and flexible enough to cover
objects and volumes beyond the reach of other robots.

ESAB North America, Inc.
1941 Heath Parkway
Fort Collins, Colorado   80522

MAC 500 ROBOTIC WELDING AND
CUTTING SYSTEM

MAC 500's easy-teach features hold training time to a minimum.
See Figure 8-14. The entire system is so simple to use that anyone can
learn its programming and operating functions in about fifteen minutes,
according to the manufacturer. Complex welding applications can be
mastered in an hour. (See welding possibilities.)

Designed as a user-friendly robot, the MAC 500 features a menu-
prompt system that provides step-by-step operating instructions in simple, noncoded English. The MAC 500 system also includes a set of
self-diagnostic messages to help eliminate programming errors.

FIGURE 8-14   The MAC 500 welding cell (courtesy of MACK
CORPORATION)

The MAC 500 is a manual teach robot that uses a point-to-point
teaching method. Its linear and circular interpolation capabilities cut
programming time and increase productivity. With its remote teaching
pendant, it allows the operator to program, fine-tune, and edit all movements of the robot.

The MAC 500 is not dedicated to just one function. It is specially
designed for GMAW applications, yet easily interfaces with FCAW,
GTAW, PAW, and PAC applications. See Spec Sheet #4.

Repeatability and Accuracy: The MAC 500 has high-speed repeatability and accuracy – from 2.4 to 2,362 inches per minute – for a
wide range of welding and cutting applications.

Precision incremental pulse encoders and advanced design of the
arm provide ± 0.004 inch repeatability. Accuracy is maintained to

± 0.008 inch. Quick operating speeds and programmable functions greatly increase versatility, productivity, and part quality. The MAC 500 robotic arc welding system is ideally suited for welding large batches (100 or more) of small, close-tolerance parts. See Figure 8–15.

The MAC 500 has experienced minimal downtime. A better than 97-percent reliability factor accounts for ESAB MAC 500's excellent record.

The System: The system consists of a 5-axes (four simultaneous axes and one independent axis) articulated SCARA robot, including manipulator, controller, and power unit. Integrated welding and/or cutting equipment includes a power source, wire feeder, welding torch, and special interface for plasma cutting. Optional accessories include manual and pneumatic turntables, a cassette recorder, a printer, extra I/O boards, and a hand teach pendant.

The following systems are available from ESAB: LAH 500 MIG Welding System; LAK 350 Pulsed MIG Welding System; LAP 500 Pulsed MIG Welding System; Sidewinder 55 Plasma Arc Cutting System.

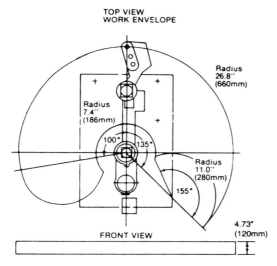

FIGURE 8–15   Work envelope for the MAC 500 robot (courtesy of MACK CORPORATION)

Feedback, Inc.
620 Springfield Avenue
Berkeley Heights, New Jersey   07922

ARMDRAULIC

Feedback's model EHA 1052A is an electrohydraulic robot designed and built for educational programs. See Figure 8–16. It has a closed-loop position control with inductive sensors. An on-board 6802 processor and a full-function teach pendant are also part of the package. Serial and parallel ports for connection to external computers are optional. Extensive documentation for service technician training is available.

ARMATROL

Feedback's model ESA 1010 is an electric servo, revolute robot designed for educational purposes. See Figure 8–17. It utilizes closed-loop position sensing that is done by using potentiometers. It is a low-cost, introductory robot. The 1010 requires an external computer. Software is available for the IBM-PC, Apple II, AIM-65, Commodore 64, and Timex 1000 computers.

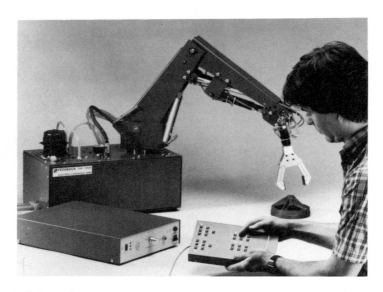

FIGURE 8–16   Feedback's EHA 1052A educational robot (courtesy of FEEDBACK, INC.)

FIGURE 8-17   Feedback's ESA 1010 computer-controlled educational robot (courtesy of FEEDBACK, INC.)

GCA Corporation
Industrial Systems Group
One Energy Center
Naperville, Illinois   60566

GCA/DKB3200 ROBOT

The GCA/DKB3200 robot is designed for heavy-duty industrial tasks such as materials handling, loading and unloading pallets, and long-reach spot welding. (See Figure 2–8.) The robot has an end-of-arm capacity of 220 pounds. It has a reach of more than 9 feet and a work envelope of about 650 cubic feet. See Figure 8–18.

All-electric DC servo-drives are on all axes, and a unique, automatic counterbalancing system is located on the arm articulation axes. This assures rapid, smooth, and responsive manipulation of maximum payloads at maximum speeds everywhere in the work envelope. The pneumatic counterbalancing facilitates the optional lead-through teaching and reduces power consumption to less than 10 percent of that required by some robots with similar payload capacities.

Up to six degrees of freedom are available with the B3200 robot. Four degrees are standard: horizontal travel (X axis), vertical travel (Z axis), horizontal sweep (theta axis), and wrist yaw (alpha axis). Wrist

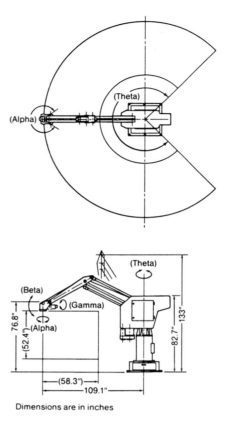

Dimensions are in inches

FIGURE 8-18   Work envelope for the GCA/DKB3200 robot
(courtesy of GCA CORPORATION/Industrial Systems Group)

pitch (beta axis) and wrist roll (gamma axis) are optional. See Spec
Sheet #5.

The B3200 robot uses a dual coordinate system (cylindrical and
Cartesian) to maximize dexterity. The robot is offered with the standard
robot controller for point-to-point applications or with the new
CIMROC 2 robot controller where either six-axes, simultaneous continuous-path, or similar integrations are required.

An externally mounted, theta-axis motor makes maintenance
easy. This motor can be changed without suspending the robot's body
by a crane or other lifting device.

The B3200 robot can be furnished with a variety of optional accessories, including mechanical gripper hands; vacuum-cup lifters for
flat material such as glass, nonferrous metal sheets, or plastic items;
magnetic lifters for iron, steel, and ferrous alloy materials; forklifts for
pallets; lifters for cartons and boxes; and complete systems for welding.

International Robomation/Intelligence
2281 Las Palmas Drive
Carlsbad, California   92008

IRI M50E AC SERVO ROBOT

Robot Command Language: The robot command language (RCL) provides the user with an easy-to-use means of creating many applications for the robot. The RCL provides a small set of easy-to-use commands, which allow the novice to create a complete application program. It also provides an extended set of commands that will satisfy the expert. The RCL is divided into the following command groups:

MOVE   SYSTEM CONTROL
TEST AND BRANCH   INPUT/OUTPUT
SET VALUE   DATA TRANSFER
DISPLAY   EDITOR
GRIPPER CONTROL   LOGIC AND ARITHMETIC

In addition to the extremely versatile RCL, the firmware provides a complete on-line editor, system control commands, and application program load and save capability. See Spec Sheet #6.

Application of the M50 Robot: The work envelope is such that it allows a wide range of activities to be performed. See Figure 8–19.

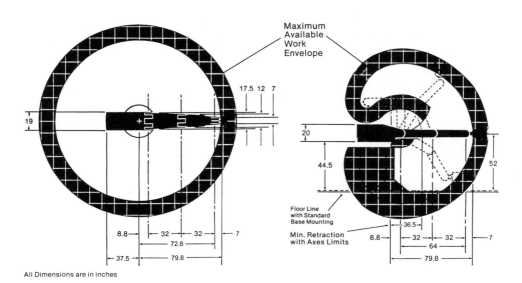

All Dimensions are in inches

FIGURE 8-19   Work envelope for the IRI M50E AC servo robot (courtesy of INTERNATIONAL ROBOMATION/INTELLIGENCE)

— Machine Loading and Unloading
   injection molding   chuckers and lathes
   compression molding   milling machines
   die casting   CNC machinery centers
   shear and punch   induction furnace
   broaching   automatic weld stations

— Palletizing and Depalletizing
   product to carton   machined parts
   carton to pallet   wood products
   food products   chemicals
   canned goods   glass
   pharmaceuticals

— Spray Applications
   mold release agent   fogging
   sand blasting   caustic washing

— Educational Application
   establishing a "hands-on" robot laboratory and control theory development

— Pick-and-Place Applications
   part transfer for final
      assembly   nondestructive testing
   test fixture load and
      unload   finished parts packaging
   conveyor-to-conveyor
      transfer   heat treating

Mack Corporation
3695 East Industrial Drive
Flagstaff, Arizona   86002-1756

MACK PRODUCTS

Mack Corporation has a line of proprietary products for the automation and automatic equipment industry. They make B·A·S·E® grippers, gripper adaptors, mounting brackets, transporters, rotators, roll models, and pitch/yaw models. They also manufacture actuators and various hydraulic and pneumatic cylinders for use in automatic equipment and robots.

Microbot, Inc.
453-H Ravendale Drive
Mountain View, California 94043

ALPHA II

The Alpha II is a proven, low-cost robot system designed specifically to help manufacturing operations improve productivity by automating low-level tasks that humans find hazardous or difficult to repeat accurately for long periods of time. See Figure 8–20.

Proven applications include machine loading and unloading, semiconductor wafer and cassette handling, solder masking, application of adhesive and coating materials, dip soldering, and packaging of pharmaceuticals and chemicals. See Figure 8–21.

This robot is a completely self-contained system that can be programmed directly with the hand-held teach control or by an external computer with an RS232 interface. The complete system consists of a five-axes articulated robot arm that can be supplied with a variety of standard or specialized grippers. The robot includes controls for two additional axes, which are available through auxiliary motors; the system control, which contains the microprocessor-based control and the computer interface and work cell interface electronics; the hand-held teach control, which makes it easy to teach the robot motions; and the opera-

FIGURE 8–20 Alpha II robot system (courtesy of MICROBOT, INC.)

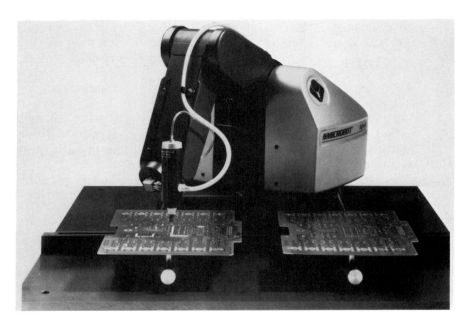

FIGURE 8–21   Alpha II dispensing system (courtesy of MI-
CROBOT, INC.)

tor control, which makes it easy for production people to select up to four separate preprogrammed operations.

     This robot can operate continuously in a broad range of production tasks, handling payloads of up to 3 pounds with repeatability requirements of up to ±0.015 inch. Maximum arm speed is 51 inches per second. See Spec Sheet #7. The semiclosed loop control system references the arm location to a home position for initialization and calibration. Six stepper motors in its base operate the robot's six-jointed arm through a system of precision stainless steel cables. Tests have shown that Alpha II has an MTBF greater than 4000 hours and an uptime of more than 98 percent.

     The removable teach control allows you to program the robot's microprocessor in real time as the robot arm is moved through each of the motions required to execute its tasks. It has an easy-to-read LED display that shows the system's status and quickly tells what step the arm is on (which allows for easy programming and editing). The microprocessor retains up to 227 individual steps. The Alpha II is also equipped with sophisticated, yet simple-to-program software that allows it to perform subroutines and loops as well.

     The Alpha II can be programmed and controlled by a host computer through its RS232C serial interface. This allows the robot to integrate with other RS232-controlled devices through the computer. Two optional auxiliary motors may be used to operate application peripherals such as rotating tables, slides, and conveyor belts, all under robot control. Eighteen optically isolated I/O ports are available for interfacing the robot with work-cell sensors and device actuators at factory-level voltages.

The operator control is used by production personnel to run pre-programmed tasks. It contains an emergency shutoff switch, but the operator cannot in any way modify or erase any of the existing programs. See Figure 8–22.

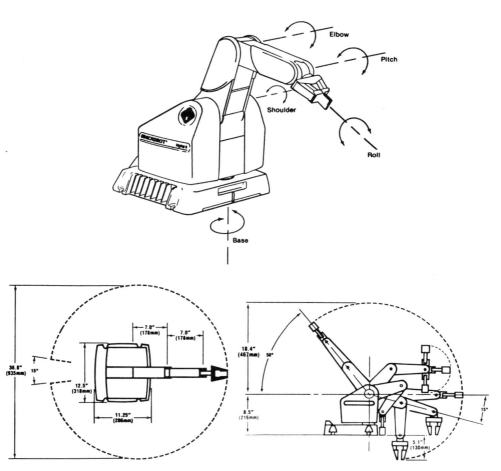

FIGURE 8–22   Operating envelope for Alpha II (courtesy of MICROBOT, INC.)

Prab Robots, Inc.
P.O. Box 2121
Kalamazoo, Michigan   49003

PRAB INDUSTRIAL ROBOTS

Many of the industrial robots built by Prab in the early 1960s are still in service today. As industry's acceptance of robots has increased, demand for additional models has likewise increased. Today, Prab offers a complete line of industrial robots including electrohydraulic and totally electric robots. See Spec Sheet #8.

The line includes a full range of cylindrical coordinate, spherical coordinate, and articulated arm robots. These robots offer point-to-point, multipoint, continuous-path, and circular interpolation capabilities.

Although Prab robots are capable of performing many different jobs, Prab has established itself as a leader in materials handling applications. Payloads of these robots range from a few ounces to one ton.

Tooling: Other robot manufacturers ship robots without end-of-arm tooling, but Prab robots are shipped with tooling installed.

Vacuum, hydraulic, electric, or magnetic gripper operations are available. Required repeatability and tolerances are factored into the design of the robot system and its tooling. See Figure 8–23.

Controllers: Prab controllers integrate state-of-the-art technology, which delivers greater robot uptime, control board axis module interchangeability, and ease of serviceability and which requires minimal floor space. The controller is designed for effective operation in a wide range of environmental conditions. Control boards are housed in the stable environment of the cabinet and have high immunity to electrical noise and temperature. Authorities consider this controller to be one of the most user friendly robot controllers on the market today.

Vision: Vision tracking in real time is available on certain Prab robots and can be interfaced with other system components such as AGVS and overhead and floor-mounted monorails.

Applications: Some of the many materials handling applications that are providing cost-effective service to customers include, but are not limited to:

high-speed machine loading/unloading. See Figure 8–24.

press-to-press transfer of parts	plastics
	glass
palletizing/depalletizing	investment casting
packaging	textiles
forging	munitions
die casting	

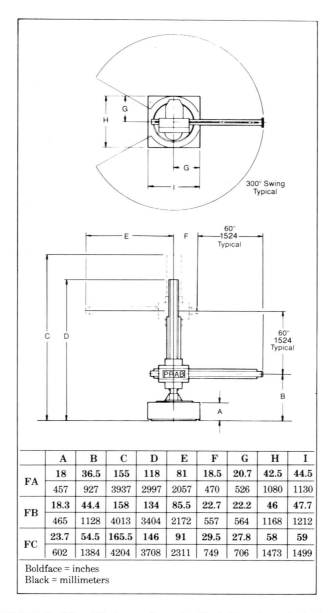

	A	B	C	D	E	F	G	H	I
FA	**18**	**36.5**	**155**	**118**	**81**	**18.5**	**20.7**	**42.5**	**44.5**
	457	927	3937	2997	2057	470	526	1080	1130
FB	**18.3**	**44.4**	**158**	**134**	**85.5**	**22.7**	**22.2**	**46**	**47.7**
	465	1128	4013	3404	2172	557	564	1168	1212
FC	**23.7**	**54.5**	**165.5**	**146**	**91**	**29.5**	**27.8**	**58**	**59**
	602	1384	4204	3708	2311	749	706	1473	1499

Boldface = inches
Black = millimeters

FIGURE 8–23   Work envelope for Prab F-series robots (courtesy of PRAB ROBOTS, INC.)

FIGURE 8-24   The Prab Model 5800 robot tending two multi-spindle machine tools in the manufacture of cylinder heads (courtesy of PRAB ROBOTS, INC.)

MODELS FA/FB/FC

Prab F-series robots are cylindrical coordinate units that deliver a smooth, steady motion for payloads of up to 2,000 pounds. They are ideally suited for heavy-duty, servo-controlled, high-speed materials handling.

In terms of weight handling, function, control operations, and up to seven-axes motion, the ultimate in flexible automation is offered. These units can be equipped with real-time vision tracking and are easily interfaced with other system components such as automatic-guided vehicles.

F-series robots provide the rigidity needed for demanding applications. A servo-feedback device on each axis is driven directly by the moving member rather than the actuator. This means accurate control and repeatability with greatly reduced backlash for long-term, close-tolerance work.

Simultaneous servo-control of up to seven axes is possible with Prab's new 700 controller, which features a 16-bit microcomputer. A free-standing, sound-enclosed, hydraulic power unit is connected to the robot by two 20-foot quick disconnect hoses as standard equipment.

Applications: F-series robots are being used effectively for the following applications:

resistance (spot) welding	palletizing
machine tool loading	glass handling
material transfer	press loading
investment casting	stacking/destacking
forging	

Schrader-Bellows
A Division of Parker-Hannifin
200 West Exchange Street
Akron, Ohio   44309-0631

ROBOTICS WITH MOTIONMATE™

MotionMate robots handle simple, repetitive materials handling tasks with ease. Transferring, loading, and unloading are simple applications, but they are tasks that can free humans for more productive use. (See Figure 1–3.) MotionMate robots are more than value-added devices; they are efficient links between different machines as well as between machines and material supply sources. They provide efficiency that pays for itself with increased production.

Typical Applications: A typical application for MotionMate robots is as simple work cells. In this application, work is first processed by one machine and then by a second machine. A robot unloads the first machine and tranfers the part to the second machine for the next operation. See Figure 8–25. This linking of operations saves labor time and reduces in-process inventories. A complete work cell must also include a method for loading the first machine and unloading the second machine. An expanded work cell links three machines and utilizes two robots.

MotionMate robots are also used in secondary operations. For example, a plastic part cannot be damaged after molding due to cosmetic requirements. The robot removes the part from the opened mold and places it in a trim die for flash removal. The finished part, with no scratches or mars, is then placed on a conveyor. See Figure 8–26.

Another application for these versatile robots is punch press loading. In Figure 8–27, a punch press is used to assemble two cup-shaped parts that form a wheel for office furniture. The robot arm automatically ejects the finished product as it moves the unfinished product into position. Even though, in this instance, the parts supply was fed from a hand-loaded magazine, the robot system tripled output because safety requirements for human punch press operators were eliminated.

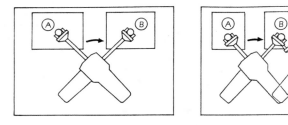

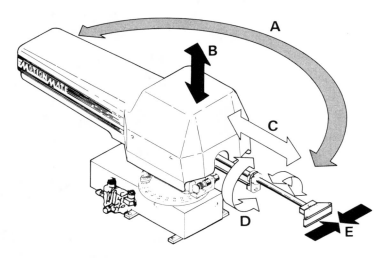

A — **Base Rotates, infinitely adjustable up to 180°.**
B — Lift of 3"
C — Extend to 12"
D — Wrist Rotate 90° or 180°
E — Grasp

FIGURE 8-25   Simple work cells—a typical application for MotionMate robots (courtesy of SCHRADER-BELLOWS)

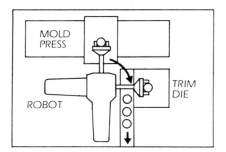

FIGURE 8-26   MotionMate robot used in a secondary operation (courtesy of SCHRADER-BELLOWS)

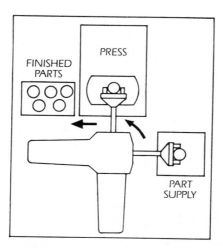

FIGURE 8–27   MotionMate robot used in a press loading application (courtesy of SCHRADER-BELLOWS)

MotionMate robots can also be used for sorting. For example, Figure 8–28 shows the robot sorting incoming parts in three categories – A, B, and C – based on the results of testing equipment. The multiple stops on the robot base and the communications linkage between the test equipment and robot are examples of custom systems work that can significantly expand the capabilities of the basic robot unit.

Rotary tables are used in many automated processes to bring parts to a variety of different work stations. In the application shown in Figure 8–29, the load robot loads parts from a delivery track, using only lift and reach movements. The unload robot, using information from the

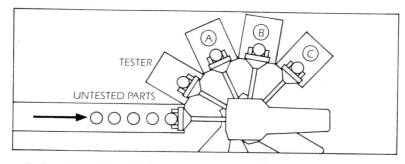

FIGURE 8–28   MotionMate robot used in a sorting application (courtesy of SCHRADER-BELLOWS)

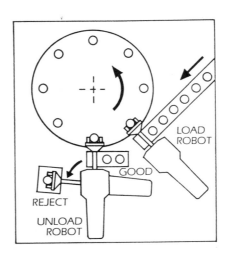

FIGURE 8-29   MotionMate robots used in loading and unloading parts from a rotary table (courtesy of SCHRADER-BELLOWS)

operations performed on the parts, places them in either good or reject bins.

The System: This highly versatile, easily programmable robot is pneumatically powered and microprocessor controlled. The MotionMate system consists of the manipulator, the robot controller, and a remote, hand-held teach module. The robot offers five axes of movement, including base rotation, lift, extension, wrist rotation, and grasp. See Spec Sheet #9.

The maximum payload is 5 pounds. One of the robot's major advantages is its ease of programming and resulting versatility. No special skills are required of the operator. The hand-held teaching module programs the robot commander, which guides the robot through a sequence of operations for quick precise transfer and placement of parts. This programming simplicity gives it the flexibility to be changed quickly for new production requirements.

Seiko Instruments USA, Inc.
Robotics/Automation Divison
2990 West Lomita Blvd.
Torrance, California   90505

CYLINDRICAL COORDINATE ROBOTS

Seiko's D-Tran RT3000-series robots are used as the standard for precision comparisons. See Figure 8-30. Accurate-positioning precision

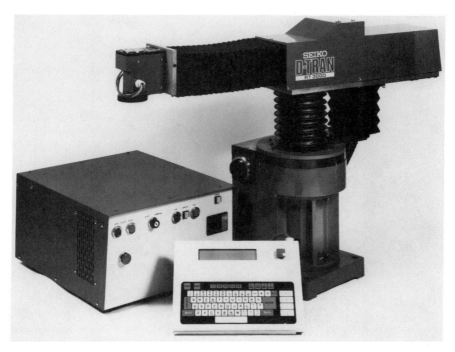

FIGURE 8–30   Seiko's D-Tran RT3000-series with power supply and programmer (courtesy of SEIKO INSTRUMENTS USA, INC.)

with high speed allows the RT-series robots to meet assembly and inspection problems with flexibility. The A axis has three possible mounting orientations: downward, upward, and outward. The R axis can be mounted in a 100 mm "extended" position. See Figure 8–31 for other details of the RT3000.

The Seiko D-Tran controller is common to all Seiko D-Tran series robots, controlling four closed-loop DC-servo axes and providing as an option a single-phase or a contamination-resistant controller.

The teach-terminal is a soft-touch, contamination-resistant, 70-key keyboard, with 40 characters by 4 lines of alphanumeric liquid crystal display (LCD).

DIMENSIONS (mm)

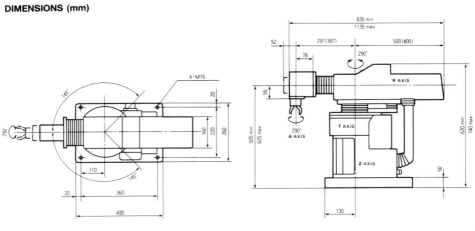

PERFORMANCE SPECIFICATIONS				OPERATING RANGE
Configuration			4 Axes	
Payload (At Maximum Speed)			5 kg (11 lbs)	
Arm Range	A:	Grip Rotation	290 deg	
	R:	Horizontal Stroke	300 mm (11.81 in.)	
	T:	Plane Rotation	290 deg	
	Z:	Vertical Stroke	120 mm (4.72 in.)	
Speed	A:	Grip Rotation	150 deg/sec	
	R:	Horizontal Stroke	750 mm (30 in.)/sec	
	T:	Plane Rotation	90 deg/sec	
	Z:	Vertical Stroke	360 mm (14 in.)/sec	
		Combined Maximum	1400 mm (55 in.)/sec	
Repeatability			±0.025 mm (0.0010 in.)	
Resolution	A:	Grip Rotation	0.005 deg	
	R:	Horizontal Stroke	0.025 mm (0.001 in.)	
	T:	Plane Rotation	0.003 deg	
	Z:	Vertical Stroke	0.012 mm (0.0005 in.)	
Weight			108 kg (238 lbs)	

● Specifications are subject to change without notice.

FIGURE 8–31   Details of the RT3000-series (courtesy of SEIKO INSTRUMENTS USA, INC.)

Unimation Incorporated
A Westinghouse Company
Shelter Rock Lane
Danbury, Connecticut   06810

UNIMATE™ Series 2000 and 4000

The Unimate Series 2000 and 4000 industrial robots are among the most widely used in the world. See Figure 8-32 and Spec Sheet #10. With over twenty years of continued improvement, they are highly reli-

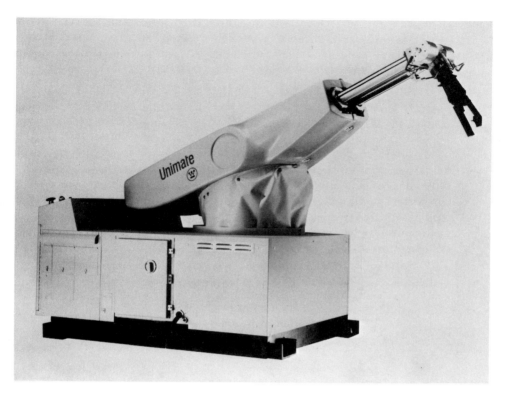

FIGURE 8-32 Unimate heavy-duty pick-and-place robot (courtesy of UNIMATION INCORPORATED, a Westinghouse Company)

able, easy-to-use robots. They have been used successfully for spot welding, die casting and investment casting, materials handling, and many other applications. See Spec Sheet #11.

The Unimate Series 2000 and 4000 robots feature six fully programmable axes of motion and are designed for high-speed handling of parts weighing up to 300 pounds. The dedicated electronic control is regarded as one of the simplest controllers available in the industry today for teaching and operating industrial robots.

UNIMATE™ SERIES 100

The Unimate Series 100 robot is designed for high-speed small parts handling in assembly, inspection, packaging, material transfer, and process applications in the automotive, appliance, electronics, aeronautical, and consumer products industries. Its high performance offers the advantages of improved production rates, consistent quality, low reject rates, and flexible operation. See Figure 8-33 and Spec Sheet #12.

PUMA® Series 200

With its high speed, repeatability, and flexibility, the PUMA Series 200 robot is suited to a wide range of small parts handling applications, and VAL control makes it easy to design application programs to carry out the most difficult robotic tasks. See Figure 8-34.

Current assembly applications include automotive instrument panels; small electric motors; printed circuit boards; and subassemblies for radios, television sets, appliances, and more. Other applications include packaging functions in the pharmaceutical, personal care, and food industries. Palletizing of small parts, inspection, and electronic parts handling in the computer, aerospace, and defense industries round out the present installed base. See Spec Sheet #13.

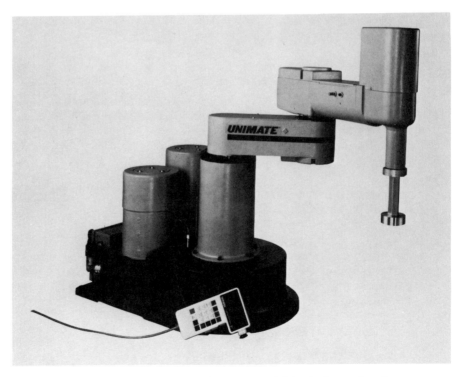

FIGURE 8-33   Unimate Series 100 pick-and-place robot (courtesy of UNIMATION INCORPORATED, a Westinghouse Company)

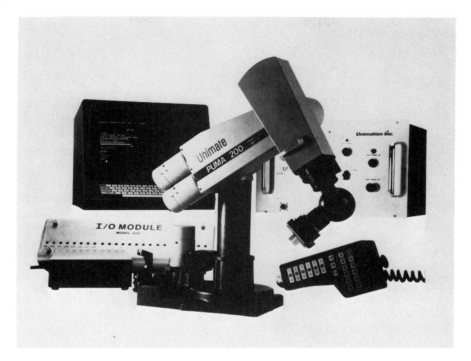

FIGURE 8–34   PUMA Series 200 with peripheral equipment (courtesy of UNIMATION INCORPORATED, a Westinghouse Company)

PUMA® Series 700

PUMA Series 700 electric robots are designed with flexibility and durability to ensure long life and optimum performance in even the harshest, most demanding manufacturing environments. See Figure 8–35. Specific customer needs for either higher payload or extended reach determine which model is suitable for a particular task. See Figure 8–36 and Spec Sheet #14.

Program: A sample pick-and-place program follows.

000 SP	2	Sets arm speed to $1/6$ of maximum speed.
001 MP	0	Moves arm to point 0 (system defined as the home position).
002 DE	10	Delays 1 second.
003 SP	4	Sets arm speed to $1/3$ of maximum speed.
004 CS	1	Branches to subroutine 1.
005 MP	2	Moves to point 2 (first PLACE point).
006 CS	2	Branches to subroutine 2.

FIGURE 8–35 PUMA Series 700 with control unit and teach pendant (courtesy of UNIMATION INCORPORATED, a Westinghouse Company)

007 CS	1	Branches to subroutine 1.
008 MP	3	Moves to point 3 (second PLACE point).
009 CS	2	Branches to subroutine 2.
010 CS	1	Branches to subroutine 1.
011 MP	4	Moves to point 4 (third PLACE point).
012 CS	2	Branches to subroutine 2.
013 GO	0	Branches to label 0 (system defined as the start of the program).
014 LB	1	Defines the beginning of subroutine 1 (PICK subroutine).
015 MP	1	Moves to point 1 (PICK point).
016 GR	0	Opens gripper.
017 GR	3	Lowers gripper.
018 WT	6	Waits for input signal to become true, then continues program.
019 GR	1	Closes gripper.

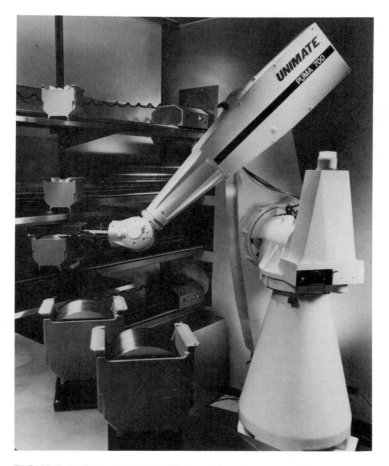

FIGURE 8-36   PUMA 700 in action (courtesy of UNIMATION INCORPORATED, a Westinghouse Company)

020 GR	2	Raises gripper.
021 WT	5	Waits for input signal to become true, then continues program.
022 RE		Returns to original program and continues.
023 LB	2	Defines the beginning of subroutine 2 (PLACE subroutine).
024 GR	3	Lowers gripper.
025 WT	6	Waits for input signal to become true, then continues program.
026 GR	0	Opens gripper.
027 GR	2	Raises gripper.
028 RE		Returns to original program and continues.

Yaskawa Electric America, Inc.
3160 MacArthur Boulevard
Northbrook, Illinois   60062

MOTOMAN® L10WA ROBOT

The Motoman L10WA robot has a maximum reach of 1,424 mm (56.06 inches). The manipulator provides 4 m³ (140 ft³) volume of working range, the largest working area for this class of industrial robot. See Figure 8–37.

The lightweight configuration of the manipulator, along with the sophisticated controller and its high-speed calculations, insures accurate, smooth, and high-speed motions over the whole range of its operation.

By using a cathode ray tube (CRT), the new RX controller displays various data such as job data, alarm contents, working time, position data, and diagnosis. See Figure 8–38.

Advanced functions such as 3-D shift, mirror image shift, arc sensor COM-ARC, and palletizing are available as options.

By utilizing 64 input and 31 output signals, 64 internal relays, and many instruction functions, sophisticated control of peripheral equipment is obtainable. By connecting the controller to a cassette tape recorder and to a printer, taught data can easily be presented and modified if necessary. The robot can be mounted the standard way, upside down, or on a wall. Manipulators are available for the various types of mounts. The main control unit can be separated from an operation panel by as much as 80 feet.

A teach-lock mode, a machine-lock function, an interference-free function, and an in-guard safety mode are built in.

Check Spec Sheets #15, #16, and #17 for details on the L10W (which is used by Hobart Brothers Company and is shown in Figure 5–17 in a welding unit) and the S50, which is shown in Figure 8–39. These are but two of the many Motoman robots from one of Japan's most versatile manufacturers of robots.

FIGURE 8-37   Motoman L10WA robot (courtesy of YASKAWA
ELECTRIC AMERICA, INC.)

FIGURE 8-38   Motoman robots with YASNAC RX controller
(courtesy of YASKAWA ELECTRIC AMERICA, INC.)

FIGURE 8-39 Motoman S50 robot (courtesy of YASKAWA ELECTRIC AMERICA, INC.)

Thermwood Robotics
P.O. Box 436
Dale, Indiana   47523

PR SERIES OF SPRAY PAINTING ROBOTS

Thermwood has been providing industrial robots and computer-controlled automation for over ten years. More than five hundred companies use Thermwood systems. The PR series boasts an impressive array of today's advanced technological features from "Path Perfect," an on-line editing system, to advanced electrical, electronic, and hydraulic diagnostic systems. See Figure 8-40. Features include a menu-driven operating system, advanced digital servos, program queue, and fine synchronization. See the work envelopes for these robots in Figure 8-41.

PR series robots are not delicate lab machines; they are designed to offer years of service to the hostile, rugged environment of a factory. Simple rugged construction borrows techniques from high-performance jet fighter aircraft to provide superb performance and extreme reliability. All electric and hydraulic systems are located in the base, eliminat-

FIGURE 8–40   Thermwood's series PR 16, 24, and 36 spray painting robots (courtesy of THERMWOOD CORPORATION ROBOTICS DIVISION)

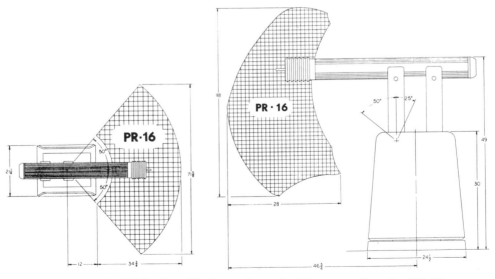

FIGURE 8–41   Work envelopes for Thermwood's series PR 16, 24, and 36 robots (Left shows top view; right, side view.) (courtesy of THERMWOOD CORPORATION ROBOTICS DIVISION)

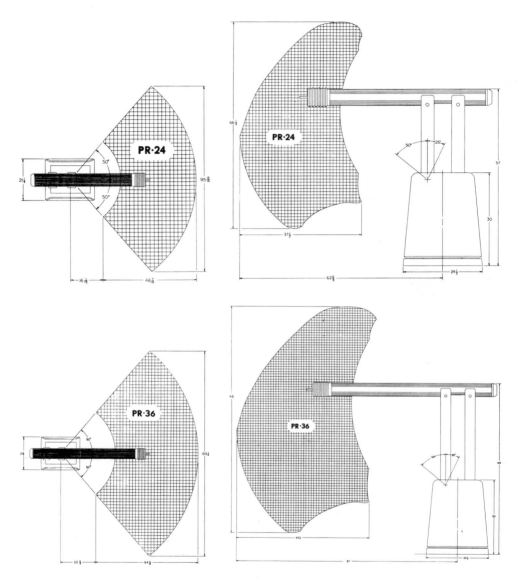

FIGURE 8-41   (Continued)

ing the problems caused by wires and hoses flexing in the arms. Oversized bearings, bushings, and push rods assure a long reliable operating life. Check Spec Sheet #18 for more details on this series of spray painting robots.

International Business Machines Corporation
Manufacturing Systems Products
P.O. Box 1328-4327
Boca Raton, Florida   33432

MANUFACTURING SYSTEMS

The IBM® 7575 and 7576 manufacturing systems are electric-drive, programmable units suitable for a wide range of industrial applications such as electronic component insertion, testing, packaging, and surface-mount device placement, as well as many light mechanical applications. Figure 8–42 shows a typically configured IBM 7575 manufacturing system, including the manipulator, a rack-mountable 7532 industrial computer model 310, a 7572 servo power module, and an AML/2 manufacturing control system licensed program. An optional hand-held, push-button pendant can be used to teach the robotic arm

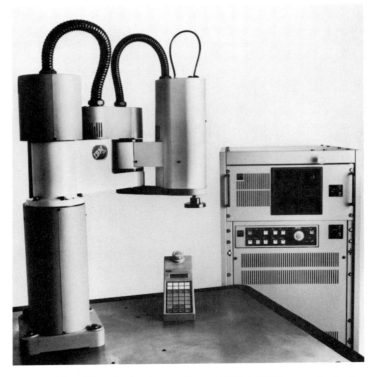

FIGURE 8–42 Typically configured IBM 7575 manufacturing system (courtesy of INTERNATIONAL BUSINESS MACHINES CORPORATION)

assembly movements. The new systems perform tasks at speeds significantly faster than IBM's current SCARA (selective compliant assembly robot arm) robotic product offerings.

Modularity and Flexibility: The rack-mountable 7532 industrial computer model 310 is the newest member of IBM's industrial computer family. The 7532 is designed to meet industrial environmental conditions such as temperature variations, vibration and shock, voltage surges, and particulates in the air. Model 310, based on the Intel 80286 microprocessor, offers the power and flexibility required for manipulator control. It has 512KB of memory and an Intel 80287 math co-processor. The computer is specially configured with two intelligent axis control cards based on the Motorola 68000 microprocessor, a 48-point digital input/digital output card, a four-port RS232 asynchronous communications adapter, and three expansion slots for optional features.

The 7572 servo power module, which can be mounted in a rack or panel, provides the power interface between the 7532 industrial computer and either the 7575 or 7576 manipulators. An easy-to-use operator control panel allows plant floor personnel to control the manufacturing system and the application program. A hand-held, push-button teach pendant is optional.

The 7575 manipulator is designed for a wide range of electronic and light mechanical assembly applications that require high speed, accuracy, and repeatability. With a maximum payload of 5 kg (11 lbs), the robotic arm can move at speeds up to 5.1 m/sec (200 in/sec) with repeatability of ± 0.025 mm (± 0.001 in), and a maximum reach of 550 mm (21.6 in). See Spec Sheet #19.

The IBM 7576 manipulator can handle assembly applications requiring heavier payloads and a larger workspace. See Spec Sheet #20. With a maximum payload of 10 kg (22 lbs) and a symmetrical workspace with a maximum reach of 800 mm (31.5 in), the manipulator is capable of speeds up to 4.4 m/sec (173 in/sec) and repeatability of ± 0.05 mm (± 0.002 in). See Figure 8–43.

New Programming Language: IBM has announced the following licensed programs for the system:

- AML/2 Manufacturing Control System, which is a prerequisite for each 7575 or 7576 manufacturing system. It contains a powerful and flexible, easy-to-use interpretive language in which robotic application programs are written.
- AML/2 Application Development Environment, which is recommended for each manufacturing site using a 7575 or 7576 manufacturing system. The program is used to easily

FIGURE 8–43   IBM 7576 manipulator (courtesy of INTERNATIONAL BUSINESS MACHINES CORPORATION)

and efficiently develop and debug customized robotic application programs.
– AML/2 Application Simulator, a powerful off-line tool for simulating and debugging AML/2 application programs. Other uses include estimating an application's cycle time and familiarizing the user with AML/2.

□ SPEC SHEETS

Load Capacity
Load 10 in (250 mm) out and 5 in (125 mm) offset
  from tool mounting plate . . . . 70 lb (32 kg)

Number of axes, configuration, servo system, control type
Number of servoed axes 6
Configuration . . . . Articulated
Drive system . . . . DC Motor
Position feedback . . . . Brushless resolver
Control type . . . . Controlled path at Tool Center Point

Positioning repeatability
Repeatability to any previously taught point . . . . ±0.010 in (±0.25 mm)

Range of motion, velocity
Base sweep . . . . 270 deg
Horizontal reach to tool mounting plate . . . . 99 in (2515 mm)
Vertical reach to tool mounting plate . . . . 137 in (3480 mm)
Pitch and Yaw relative to forearm . . . . 238 deg
Roll relative to wrist . . . . 900 deg
Nominal velocity at Tool Center Point (TCP) . . . . 25 ips (635 mm/sec)
Base slew rate . . . . 95 deg/sec

Memory capacity, I/O contacts
Data area size/Average number of data points . . . . 48 k byte/3000
Number of input contacts/Maximum optional 16/32
Number of output contacts/Maximum optional 16/32

Floor space and approximate net weight
Robot . . . . 15ft² (1.4 m²)/5,250 lb (2385 kg)
Robot control . . . . 8.1ft² (0.75 m²)/1,470 lb (670 kg)

Ambient temperature . . . . 40 to 105°F (5 to 40°C)
  with air conditioner option . . . . 40 to 120°F (5 to 50°C)

Power requirements . . . . 460 volt, 3φ, 60 Hz*
Power rating/Power required for typical cycle 23kVA/4kW

*Other voltages and 50 Hz available

Spec Sheet #1   T³746 industrial robot specifications (courtesy
of CINCINNATI MILACRON)

SMART 6.50 GENERAL DATA		
Robot Type	All-electric, articulated anthropomorphic configuration	
Number of Axes	6 axes (1 or 1T, 2, 3, 4, 5, 6)	
Axes Motion Range and Speed — 1-Axis	Base Rotation: 270° (78°/sec)	
1T-Axis	Base Traverse: 2,5 m (or multiples) (1 m/sec)	Option
2-Axis	Arm Motion: 140° (80°/sec)	
3-Axis	Fore-Arm Motion: 280° (102°/sec)	
4-Axis	Fore-Arm Roll: 360° (138°/sec)	
5-Axis	Yaw/Pitch: 230° (136°/sec)	
6-Axis	Tool Roll: 360° (144°/sec)	
Reach	Horizontal 1917 mm, vertical 3000 mm	
Repeatability	± 0.4 mm	
Base Floor Space	1150 × 1150 mm	
Load Capacity — Static Load	50 kg (400 mm from wrist center), or 60 kg (330 mm)	
Moment of Inertia	5-Axis: 8 kgm² - 6-Axis: 1.125 kgm²	
Static Torque	5-Axis: 200 Nm - 6-Axis: 75 Nm	
Drive	Electric DC servo motors with transistor PWM amplifiers	
Position Transducers	3 KHz resolvers mounted on motor shafts	
Axes Counterbalancing (1)	Pneumatic for the 2-Axis and weight for the 3-Axis (floor and roof mounting only)	
Axes Counterbalancing (2)	Weight for the 2 and 3 Axes (universal position mounting)	Option
Safety Flange	Available (for wrist overload protection)	Option
Axes Travel Limits	Programmable software limits	
	Adjustable limit switches for 1, 2 and 3 axes	Option
	Electrical limit switche for 4 axes (fixed)	
	Energy absorbing stops for 1, 2, 3, 4, 5 axes (1 and 3 may be adjusted)	
Robot Fine Calibration Device	Available	Option
Pins kit for Robot Lifting and Rotation	Available	Option
Provision for Fork-Lift Hoisting	Standard	
Computerized Control System	Multi-microcomputer configuration	
Serial Interface	Available for host computer and/or sensor subsystems connection (RS232 or RS422)	Option
Diagnostics	Wide range of controller diagnostic functions	
Controller Size	1250 × 1580 × 930 mm	
Controller Doors Interlocks	Available	Option
Program Storage	RAM CMOS with battery back-up (minimum 50-days retentivity)	
Program Storage Capacity	16, 32, 48, or 64K bytes. Number of programs limited only by memory capacity	16K standard
I/O Interface	14 Inputs, 9 Outputs user programmable standard	
I/O Interface Extension	5 additional I/O modules available (32 Inputs or 16 Outputs per module)	Option
Back-Up Recording Unit	Portable cassette tape unit	Option
Printer for Program Listing	Available	Option
Operator's Panel	9" CRT, Alphanumeric Keyboard, and Control Panel	
Control Panel	Electronics and Motor Drives ON/OFF, Emergency Stop, Feed Hold, Axes Zero, Feed Rate Selector, Program Execution Mode Selector, Cycle Start	
Programming Terminal (Pendant)	Microcomputer based. Includes key-pad, display, emergency stop, and dead-man switch	Option
Emergency Manual Control Box	Available	Option
Programming Method (1)	Teaching through the Programming Terminal (motion, interlocks, tool functions)	
Programming Method (2)	Manual Data Input (MDI) through the Operator's Panel using the PDL Programming System	
Programming Method (3)	Edit/Teach by MDI and teaching of key-positions	
Motion Mode (1)	Point-to-Point in the Robot Coordinate System	
Motion Mode (2)	Point-to-Point with path control (straight and circular line) of the tool centre point (TCP) in the Absolute Cartesian Coordinate System	Option
Motion Mode (3)	Point-to-Point with path control of tool centre point (TCP) in the Tool Cartesian System	Option
Selection of Motion Mode	From Program, Programming Terminal, and Operator's Panel	
Program Execution	Foreward and Backward Jogging, Automatic, Step-by-Step, Continuous Repeated	
Power Requirements	380 V three-phase + 10%, − 15%, (or others on request), 48-62 Hz, 14 KVA	
Air Supply	6 bar	
Environmental Operating Range	0–45 °C	
Robot Weight	1600 kg (with pneumatic balancing)	

Spec Sheet #2   SMART 6.50 general data (courtesy of COMAU PRODUCTIVITY SYSTEMS, INC.)

1 (x)[1]	T up to 7.5m (24.6 ft.)	1 m/sec. (3.3 ft./sec.)	80 kg (175 lb.)
3 (z)[1]	T up to 1.5m (4.9 ft.)	0.5 m/sec. (1.65 ft./sec.)	80 kg (175 lb.)
5	R ± 105°	3.0 rad./sec.	20 m-kg (145 ft.-lb.)[2]

Specifications are subject to change without notice.   [1]These dimensions can be exceeded to fit a wide range of requirements.
[2]Allowable torque load perpendicular to axis of rotation

Drive System
Electric or hydraulic motors with with rack and pinion drives and gear reduction

Accuracy
With 80 kg load at maximum speed

Position Accuracy: ± 0.5mm (0.020″)
Position Repeatability: ± 0.2mm (0.008″)

Control System
The RC-6 is the standard Cybotech controller, specifically designed for ease of operation, flexibility, and simple maintenance. The RC-6 incorporates diagnostic circuits that greatly reduce troubleshooting time, while providing many safety features. Refer to RC-6 brochure for more details.

Power Requirements
20 kVA, 440 VAC, 3 phase (x2 for double G80) total for power unit and control system

Cooling Water (Hydraulic only)
11 gpm max. (x2 for double G80)

Weight
4,200 - 12,000 kg (9,000 - 26,000 lb.) depending on gantry dimensions and single or double configuration

Spec Sheet #3   G80 industrial robot engineering data (courtesy of CYBOTECH INDUSTRIAL ROBOTS)

MAC 500 MANIPULATOR

Structure	Horizontally Articulated
Axes of Motion	4 Simultaneous – 1 Independent
Load Capacity	6.6 lbs. (3 kg) Maximum
Position Repeatability	±0.004″ (±.1mm) Maximum
Position Accuracy	±0.008″ (±.2mm) Maximum
Weight–Manipulator	120 lbs. (54 kg)
Base	154 lbs. (70 kg)

OPERATING RANGE OF EACH AXIS

	Stroke	Maximum Speed
(1) First Arm	0-235°	143°/sec
(2) Second Arm	0-155°	204°/sec
(Z) Vertical Wrist	4.73″ (120mm)	7.9″/sec (200mm)
(R) Turning Wrist	0-380°	150°/sec
(T) Torch Angle Wrist	0-180°	100°/sec

RANGE OF EFFECTIVE WORKING SPEED

2.4 IPM (0.06 m) to 2362 IPM (60 m)

MAC 500 CONTROLLER AND POWER UNIT

Weight–Control & Power Unit	319 lbs. (145 kg)
Teaching Method	"Direct" manual teaching MDI
	Teaching Box Operation (Optional)
Path Control	Continuous Path Control by PTP teaching
Interpolation	Linear or Circular
Number of Controlled Axes	4 Simultaneous Axes T Axis is independently controlled
Position Control Position Detection Method	Digital Closed Loop
	Pulse encoder, orgin position alignment
Speed Control	Constant Linear Speed Control
Memory/Memory Capacity	IC Memory/16K 500 steps/500 sequences
Battery Backup	Two Weeks (Rechargable)
Number of Programs	8
Program Size	255 Steps
Display	9 inch CRT

OPTIONAL EQUIPMENT

Teaching Box	16½′ (5m) of Cable
Extended I/O Board	Input - 8 Terminals (for interlock)
	Output - 8 Terminals (for synchronizing)
Audio Cassette Tape Recorder Interface	External Memory for Data Saving and Data Loading
Printer Unit	Electrical Discharging Type Dot-Matrix

Spec Sheet #4   MAC 500 specifications (courtesy of MACK CORPORATION)

Model	DKB3200 robot
Degrees of Freedom	6 (Beta and Gamma optional)
Payload, max including end effector weight	220 lb
Repeatability	±0.04 in.

Range (speed)	Axes	
	X (Horizontal, in and out)	58.3 in. (35.4 ips)
	Z (Vertical travel)	52.4 in. (35.4 ips)
	Theta$_1$ (Horizontal swing)	270° (75°/sec)
	Theta$_2$ (Main arm vertical rotation)	—
	Theta$_3$ (Forearm vertical rotation)	—
	Alpha (Wrist yaw)	300° (75°/sec)
	Beta (Wrist pitch)	120° (75°/sec)
	Gamma (Wrist rotation)	180° (75°/sec)
	Composite axes speed	—

Robot weight	3300 lb
Electric power requirement	AC 200/220v, 50/60 Hz, 2.5 KVA
Pneumatic counterbalancing (where applicable)	Pneumatic counterbalancing on X and Z axes

—All **GCA**/DK pedestal robots have a DC servomotor drive system
—Pneumatic power requirement (where applicable) is 85 psi.

Spec Sheet #5   GCA pedestal model robot specifications (courtesy of GCA CORPORATION/Industrial Systems Group)

Item   Specification

Robot Arm

Axis	3 Axis, Torso, Shoulder, Elbow standard
Clearance Required	Sphere, 80-inch radius (excluding end-effector)
Weight	350 lbs. (without optional base)
Drive	Digital AC Servo Motor
	Three Phase AC Induction Motor
Maximum Load	50 lbs. at the face plate (including end-effector)
Reach	Minimum 20 inches/Maximum 80 inches
Position Repeatability	± .040 inches (± 1 mm)

Motion, Range & Maximum Speed

Torso	360° (1½ revolutions) 60° per second
Shoulder	180° (up 90°, down 90°) 30° per second
Elbow	230° (up 90°, down 140°) 45° per second
Wrist-pitch * (See options below)	240° (± 120°) 120° per second
Wrist-roll * (See options below)	360° (2 revolutions) 120° per second

Control

Microcomputer	Master Robot Control System
(Torso mounted)	(Motorola 68000 w/32KB Memory Computer)
	3 Independent Axis Control System Computers
	(2 Optional)
	3 AC Servo Output Drive Computers
	(2 Optional)
	Dedicated Safety Computer
Input/Output Ports	
	8 bit input/8 bit output (optional 16/16)
Communications	
	2 Independent RS232C Connections

Programming

Language	IRI Robot Command Language (RCL)
Type	Basic-like language structure
Development	On Line Interpretive
	On line editor and teach-playback

Environment

Ambient Temperature	50°C (122°F) Max.
Relative Humidity	90% Max. — non-condensing

Power Inputs

Electrical	208 VAC, 3 phase, 60Hz-4 wire-30A
Air	80-100 psi, 20-130 scfm (depends on application)
	½ inch supply line

Options

* 2 Axis Wrist-Pitch & Roll	
Base	48 inches high, 24 × 24 inch base
Universal End-effector	See Gripper Product Bulletins
Memory	Up to 128 KB
Move Pendant	
System Terminal	Keyboard, hard copy printer
Load Terminal	5¼" Floppy Disk
Utility I/O Module	8 digital in, 8 digital out (16/16 optional), E-Stop Switch
Base Installation Kit	
Pneumatic Installation Kit	FRL-Filter, Regulator, Lubricator
Back-up Battery	Supply to RAM memory
Power Input	220 VAC, 3 phase, 50Hz-4wire-30A

(Dimensions and specifications subject to change.)

Spec Sheet #6   IRI M50E AC servo robot specifications (courtesy of INTERNATIONAL ROBOMATION/INTELLIGENCE)

Description:
Robotic System with five-axis articulated arm plus controls for the cable-operated
mechanical gripper and 2 additional axes (through optional auxiliary motors.)

Performance:
Maximum Payload: 3 lb. (1.362 kg) including gripper.

Maximum Speed at End of Arm: 51 inches/sec (1295mm/sec) with up to a 1.0 lb.
(454 gm) payload. Maximum speed at 3 lb. (1.362kg) payload is 6 inches/sec.

Reach: 18.4″ (467mm) spherical radius.

Repeatability: ± 0.015″ (.38mm) with total payload up to 1.5 lb. (681 gm).

Axes: 5 axes plus controls for the mechanical gripper and 2 additional axes (through
optional auxiliary motors.)

Motion Range & Speed:	Base rotation	345°	106°/sec
	Shoulder	145°	119°/sec
	Elbow	135°	202°/sec
	Pitch	180°	315°/sec
	Roll	540°	315°/sec

Control:
Teaching Methods: Teach Control with LED display & multifunction key pad or
                      host computer.

Workcell Interface: Eighteen channels for optically isolated relay input/output at
                      factory level voltages.

Computer Interface: Dual RS-232C asynchronous serial communications inter-
                      face. Operates with any host computer language supporting
                      ASCII I/O.

Control Method: Semi-closed loop control with automatic homing.

Drive: Electrical stepper motors—half step pulses.

Arm movement: Coordinated movement—all joints move simultaneously.
                      Point-to-point or integrated path moves.

Programming Capacity: 227 program steps in non-volatile memory.

Controller: Real time 6502A microprocessor controls all functions.

Installation:
Operating Environment: Ambient temperature limits: 55 °F to
                       104 °F(13 °C to 40 °C).

Mounting: The robot may be mounted and operated right-side up
                      or upside down with the base in a horizontal plane.

Power Requirements: 115VAC @ 1.5A; 220VAC @ .75A; 100VAC @ 1.7A;
                      47-400Hz.

Weight: Robot Arm: 30 lbs. (13.6 kg)
                      System Control: 30 lbs. (13.6 kg)
                      Total Shipping Weight: 75 lbs. (34.1 kg)

This data is based on laboratory tests conducted by Microbot. Specifications are subject
to change without notice.

Spec Sheet #7   Alpha II system specifications (courtesy of MI-
CROBOT, INC.)

Technical Data	FA	FB	FC
Axes of motion (standard)	3	3	3
Payload capacity	250 lbs. (114 kg)	600 lbs. (273 kg)	2,000 lbs. (909 kg)
Repeatability	±.050" (±1.27 mm)	±.050" (±1.27 mm)	±.080" (±2.0 mm)
Motion			
Horizontal swing	300°	300°	300°
Vertical stroke			
• Standard	60" (1524 mm)	60" (1524 mm)	60" (1524 mm)
• Optional	48" (1219 mm)	48" (1219 mm)	48" (1219 mm)
Extend/retract stroke			
• Standard	60" (1524 mm)	60" (1524 mm)	60" (1524 mm)
• Optional	48" (1219 mm)	48" (1219 mm)	48" (1219 mm)
Optional wrist axes	Up to 3	Up to 3	Up to 3
Optional traverse base	Yes	Yes	Yes
Coordinate system	Cylindrical	Cylindrical	Cylindrical
Weight of robot	2,400 lbs. (1090 kg)	4,000 lbs. (1815 kg)	6,000 lbs. (2720 kg)
Hydraulic power unit (remote with 20' hose standard)	1,000 lbs. (454 kg)	1,350 lbs. (613 kg)	1,500 lbs. (680 kg)
Maximum horizontal reach (see chart 3 axis robot)	78.5" (1994 mm)	82" (2083 mm)	89.5" (2273 mm)
Servo/non-servo	Servo	Servo	Servo
Drive	Electro-hydraulic	Electro-hydraulic	Electro-hydraulic
Feedback	Absolute Resolver	Absolute Resolver	Absolute Resolver
Ambient temperatures for Robot and Hydraulic Power Unit			
• Standard water cooled	40°F-120°F (4°C-49°C)	40°F-120°F (4°C-49°C)	40°F-120°F (4°C-49°C)
• Optional air cooled	40°F-100°F (4°C-38°C)	40°F-120°F (4°C-38°C)	40°F-100°F (4°C-38°C)
• Optional refrigeration cooled	40°F-120°F (4°C-49°C)	40°F-120°F (4°C-49°C)	40°F-120°F (4°C-49°C)
Floor space required for mounting base	13 sq. feet (1.2 sq. m)	15 sq. feet (1.4 sq. m)	24 sq. feet (2.2 sq. m)

Note: Reach, speed, acceleration and deceleration all affect the above payloads and repeatability. Please consult Prab Robots, Inc., Kalamazoo, Michigan, for further information regarding your application. Payload calculation must include gripper, additional wrist axes (if required) and part weight.

Spec Sheet #8   Prab models FA/FB/FC specifications (courtesy of PRAB ROBOTS, INC.)

AXIS OF MOTION					
AXIS	DESCRIPTION	FUNCTION	STROKE	MOTION/SPEED	AXIS DRIVE
A	Base rotate	Horizontal rotation	180°	180° per second	Rack and pinion, powered by dual air cylinders
B	Lift	Vertical linear	3"	12 to 24 i.p.s depending upon load	Air cylinder
C	Extend	Horizontal linear	12"	12 to 24 i.p.s depending upon load	Air cylinder
D	Wrist	Wrist rotation	90° 180°	¼ second ½ second	Cam actuated
E	Grasp	Open/close	+18°	⅛ second	Air cylinder actuated draw bar

Weight of robot	130 pounds
Max. load capacity at wrist	5 pounds
Repeatability	+ .005
Axis drive method	Pneumatic 60 to 120 p.s.i.
Power source	115V/60Hz, 240V/50Hz optional
Axis motion detectors	Proximity switches on base rotate, lift and extend
Adjustable mechanical stops	On base rotate, lift and extend
Operating temperature	+20°F to +140°F
CONTROL	Electronic, microprocessor with remote portable teaching module
Memory capacity	Up to 300 steps
Program method	Push button via portable teach module
Maximum time interval between steps	10.79 minutes
Auxiliary equipment inputs/outputs	6 outputs and 8 inputs each rated at 115V/60Hz Other voltages available

TORQUE DEVELOPED AT WRIST	90° WRIST ROTATION		180° WRIST ROTATION	
	P.S.I.G.	INCH LBS	P.S.I.G.	INCH LBS
	60	38	60	19
	80	54	80	27
	100	70	100	35

GRIPPER CLOSER FORCES	LINEAR		PIVOT (AT END OF 2" FINGER TOOLING)	
	P.S.I.G.	FORCE IN LBS	P.S.I.G.	FORCE IN LBS
	60	19	60	20
	80	26	80	27
	100	33	100	34

Spec Sheet #9   MotionMate robot technical specifications (courtesy of SCHRADER-BELLOWS)

TYPICAL PRODUCTION ENVIRONMENT STRUCTURES

BASIC CONFIGURATION

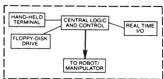

STANDARD CONFIGURATION

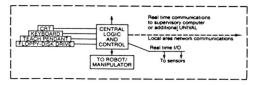

UNIVAL SYSTEM PERIPHERALS

System Terminal. Comprised of the CRT and keyboard. Program and diagnostic information are displayed via the CRT. Keyboard offers direct user interface for generating and modifying VAL III programs, as well as capabilities to initiate functions and recall stored data.

Teach Pendant. A microprocessor-based device used as a programming aid to teach location points and to move the robot arm. An alphanumeric display can show error messages and the names of previously stored locations. Individual push buttons provide manipulation of the robot arm in Joint, World, Tool, and Workpiece modes. While in the teach mode, the user gains total control over robot motion, coordinating all axes for movement along a straight line, and the ability to maintain a desired tool orientation relative to the workpiece.

Hand-Held Terminal. A computerized device used as a simplified operator interface for the BASIC configuration. It has facilities to: position the robot arm in Joint, World, or Tool modes: create and execute programs and subprograms; store and load programs on the floppy disk; and program I/O signals. The alphanumeric display informs the operator of program status and prompts the operator through the programming sequence. Simple push-button programming provides fast set-up without requiring programming experience.

Floppy Disk Drive. Provides a high-density, quick-access storage medium for programs and robot locations. It is a 1 megabyte, double-sided, double density diskette system which operates at 9600 baud and can store up to 10,000 program steps. It offers reliable program back-up, and allows fast program and data file referencing through VAL III commands.

SPECIFICATIONS

	BASIC	STANDARD
Central Processor	Motorola 68000	Motorola 68000
Language	(not applicable)	VAL III
Coordinate Frames	Joint, World, Tool	Joint, World, Tool, Workpiece
Input Lines	16 (24 VDC) or 16 (110 VAC)	32 (24 VDC) or 32 (110 VAC)
Output Lines	16 (contact closure)	32 (contact closure)
Memory Capacity	16K Bytes (CMOS) 256K Bytes (Optional)	256K Bytes (CMOS) 512K Bytes (Optional)
Memory Battery Life	30 days	30 days
Program Storage	High density floppy disk	High density floppy disk
Controller Size	49" × 31" × 26" (1245mm × 788mm × 660mm)	60" × 31" × 26" (1524mm × 788mm × 660mm) incl. CRT
Controller Weight	550 lbs. (250kg)	550 lbs. (250kg)
Number of Subroutines	Unlimited	Unlimited
Number of Nested/ Subroutine Calls	4	10
Number of Coordinate Transformations	4	Unlimited
Max. Number of Stored Points	300 (6000 Optional)	6000 (12000 Optional)
Number of Comm. Ports	1 (RS422)	1 (RS422; 4 or 8 Optional)

Spec Sheet #10   Typical production environment structures, peripherals, and specifications for Unival robot system (courtesy of UNIMATION INCORPORATED, a Westinghouse Company)

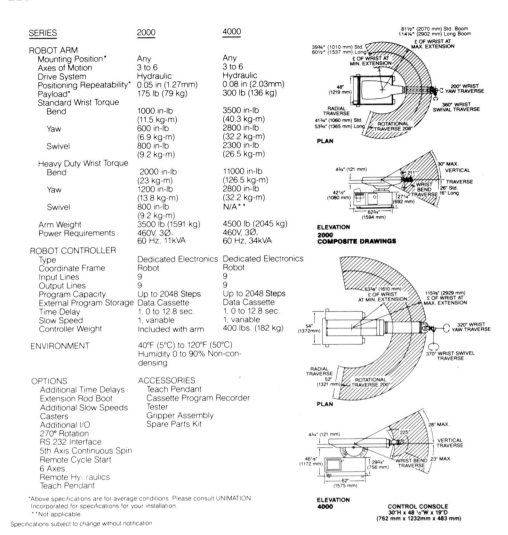

SERIES	2000	4000
ROBOT ARM		
Mounting Position*	Any	Any
Axes of Motion	3 to 6	3 to 6
Drive System	Hydraulic	Hydraulic
Positioning Repeatability*	0.05 in (1.27mm)	0.08 in (2.03mm)
Payload*	175 lb (79 kg)	300 lb (136 kg)
Standard Wrist Torque		
Bend	1000 in-lb	3500 in-lb
	(11.5 kg-m)	(40.3 kg-m)
Yaw	600 in-lb	2800 in-lb
	(6.9 kg-m)	(32.2 kg-m)
Swivel	800 in-lb	2300 in-lb
	(9.2 kg-m)	(26.5 kg-m)
Heavy Duty Wrist Torque		
Bend	2000 in-lb	11000 in-lb
	(23 kg-m)	(126.5 kg-m)
Yaw	1200 in-lb	2800 in-lb
	(13.8 kg-m)	(32.2 kg-m)
Swivel	800 in-lb	N/A**
	(9.2 kg-m)	
Arm Weight	3500 lb (1591 kg)	4500 lb (2045 kg)
Power Requirements	460V, 3Ø,	460V, 3Ø,
	60 Hz, 11kVA	60 Hz, 34kVA
ROBOT CONTROLLER		
Type	Dedicated Electronics	Dedicated Electronics
Coordinate Frame	Robot	Robot
Input Lines	9	9
Output Lines	9	9
Program Capacity	Up to 2048 Steps	Up to 2048 Steps
External Program Storage	Data Cassette	Data Cassette
Time Delay	1, 0 to 12.8 sec.	1, 0 to 12.8 sec.
Slow Speed	1, variable	1, variable
Controller Weight	Included with arm	400 lbs. (182 kg)
ENVIRONMENT	40°F (5°C) to 120°F (50°C)	
	Humidity 0 to 90% Non-con-	
	densing	

OPTIONS
  Additional Time Delays
  Extension Rod Boot
  Additional Slow Speeds
  Casters
  Additional I/O
  270° Rotation
  RS 232 Interface
  5th Axis Continuous Spin
  Remote Cycle Start
  6 Axes
  Remote Hydraulics
  Teach Pendant

ACCESSORIES
  Teach Pendant
  Cassette Program Recorder
  Tester
  Gripper Assembly
  Spare Parts Kit

*Above specifications are for average conditions. Please consult UNIMATION
Incorporated for specifications for your installation.
**Not applicable.

Specifications subject to change without notification.

Spec Sheet #11   Unimate Series 2000 and 4000 specifications (courtesy of UNIMATION INCORPORATED, a Westinghouse Company)

GENERAL
Configuration   3½ axes, horizontal plane

Drive   DC servo on the primary and
                         secondary arms and wrist.
                         Pneumatic Z (insertion) axis.

Teaching Method   Teach pendant
Program Language   C/ROS
Program Capacity   500 program steps with up to 94
                         programmable points in space

External Program
  Storage   Digital cassette recorder (option)
Gripper Control   Pneumatic
Power
  Requirements   120 VAC, 60Hz, 20a
Optional
  Accessories   (refer to page 2)

PERFORMANCE
Repeatability*   ±0.002 in. (0.05mm)
Maximum Payload   11 lbs. (5 kg)
Vertical Spring
  Rate (Arm)*   3200 lb./in. (57 kg/mm)
Radial Extension
  —Maximum*   24 in. (609.6mm)
  —Minimum   5.75 in. (146.1mm)
Velocity—linear*   150 in./sec. (3810mm/sec.)
Travel
  —Primary and
   Secondary Arms   320°
  —Wrist   Continuous (pneumatic lines
                         limit the number of possible
                         revolutions)
  —Insertion   4.5 in. (114.3mm)
Velocity
  —Primary and
   Secondary Arms   250°/sec.
  —Wrist   180°/sec.
Acceleration
  —Primary and
   Secondary Arms   1,700°/sec./sec.
  —Wrist   850°/sec./sec.
*measured with arm at full extension

ENVIRONMENTAL   50 - 122°F (10 - 50°C)
OPERATING RANGE   10 - 90% RH (non-condensing)

PHYSICAL CHARACTERISTICS
Arm Weight   160 lb. (72.7 kg)
Control Cabinet Size   32 in. × 48 in. × 12 in.
                       (813mm × 1,219mm × 305mm)
Control Cabinet
  Weight   120 lb. (54.5 kg)
Controller Cable
  Length   15 ft.
Teach Pendant
  Cable Length   15 ft.
Specifications subject to change without prior notification

Spec Sheet #12   Unimate Series 100 specifications (courtesy
of UNIMATION INCORPORATED, a Westinghouse Company)

GENERAL
Configuration   6 degrees of motion
Drive   Electric DC servos
Controller   System computer (LSI-11)
Teaching Method   By teach control and/or computer terminal

Program
Language   VAL PLUS or VAL II
Program Capacity   8K CMOS user memory in VAL PLUS
24K CMOS user memory in VAL II
Options for add'l. user memory

External Program
Storage   Floppy-disk
Gripper Control   4-way pneumatic solenoid
Power Requirement   110-130 VAC, 50-60 Hz, 500 watts

Optional
Accessories   TTY terminal, I/O module (8 input/8 output signals isolated AC/DC levels) up to 32 I/O capacity, pneumatic gripper without fingers, special software packages

PERFORMANCE
Repeatability   ±0.002 in. (0.05 mm)

Straight Line
Velocity   49 in/sec. max. (1.245m/sec.)

Maximum Payload
Static Load   2.2 lbs (1.0 kg)

Dynamic Load   24 lb-in^2 (70.4 kg-cm^2) (a 2.2 lb
Around Joint 5   (1kg) concentrated load at 3.3 in. (8.4 cm) from Joint 5)

Dynamic Load   9.0 lb-in^2 (26.4 kg-cm^2) (a 2.2 lb
Around Joint 6   (1 kg concentrated load at 2.0 in. (5 cm) from Joint 6)

ENVIRONMENTAL
OPERATING RANGE
50-120°F (10-50°C)
10-80% relative humidity (non-condensing)
Shielded against industrial line fluctuations and human electrostatic discharge

PHYSICAL
CHARACTERISTICS
Arm Weight   29 lbs. (13.2 kg)
Controller Size   12.5″ H × 17.5″ W × 19.6″ D. (317.5 mm H × 444.5 mm W × 500.0 mm D) (19 in. rack mountable)

Controller Weight   80 lbs. (36.4 kg)
Controller Cable
Length   15 ft. (4.57m) std
50 ft. (15.24m) max.

Specifications subject to change without prior notification

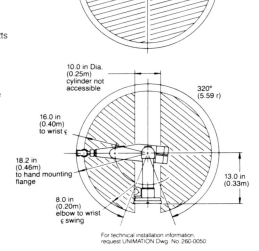

NOTE:
This region is attainable by robot in lefty configuration.
308°

10.0 in Dia.
(0.25m)
cylinder not accessible

320°
(5.59 r)

16.0 in
(0.40m)
to wrist ¢

18.2 in
(0.46m)
to hand mounting flange

13.0 in
(0.33m)

8.0 in
(0.20m)
elbow to wrist
¢ swing

For technical installation information, request UNIMATION Dwg. No. 260-0050

Spec Sheet #13   Puma Series 200 specifications (courtesy of UNIMATION INCORPORATED, a Westinghouse Company)

GENERAL
- Configuration — 6 degrees of motion
- Drive — Electric DC Servos
- Controller — System computer (LSI-11), with CRT
- Teaching Method — By teach pendant and/or computer terminal
- Program Language — VAL II
- Program Capacity — 24KW CMOS user memory in VAL II
- External Program Storage — Floppy Disk (Double sided, double density)
- I/O Module — 32 input/32 output AC/DC user selectable voltage range, plus control I/O s
- Gripper Control — 4-way pneumatic solenoid
- Power Requirement — 240/380/415/480 Volts, 3-phase, 4500 watts (peak)

ENVIRONMENTAL OPERATING RANGE
50°-120°F (10°-50°C)
95% relative humidity (non-condensing)
Shielded against momentary industrial line fluctuations of -15%, +200% of nominal voltage and up to 10KV electrostatic discharge. All covers and rotary seals have gaskets to protect against water spray and dust.
Designed to comply with IP54 and NEMA 12 International Electrical Packaging Specifications.

PHYSICAL CHARACTERISTICS

	Model 761	**Model 762**
Arm Weight	1276 lb. (580kg)	1298 lb. (590kg)
Base Diameter	23.6 in. (0.6m)	23.6 in. (0.6m)
Control Cabinet: Size	45.6 in.H x 23.6 in.W x 31.5 in.D (1160mm x 600mm x 800mm)	
Weight	440 lb. (200kg)	

PERFORMANCE

	Model 761	**Model 762**
Repeatability	±0.008 in. (0.2mm)	±0.008 in. (0.2mm)
Maximum Payload Static Load	22 lb. (10kg)	44 lb. (20kg)
Dynamic Load Around Joint 5	1365 lb.-in.^{2} (4000kg-cm^2) 10kg load concentrated 20cm from JT5 (tool flange is 12.5cm from JT5)	4270 lb.-in.^{2} (12500kg-cm^2) 20kg load concentrated 25cm from JT5 (tool flange is 12.5cm from JT5)
Dynamic Load Around Joint 6	340 lb.-in.^{2} (1000kg-cm^2) 10kg concentrated at 10cm from JT6.	683 lb.-in.^{2} (2000kg-cm^2) 20kg concentrated at 10cm from JT6.
Straight Line Velocity	40 in./sec. (1.0m/sec.) max.	40 in./sec. (1.0m/sec.) max.

Specifications subject to change without prior notification

Model 762

320°

(Model 761—1500mm)

1250

11.75°

(Model 761—13.1°)

Inaccessible area (can be reached in lefty configuration)

Robot shown in righty configuration

Cylindrical volume 112mm diameter inaccessible to JT6 tool flange

514mm radius Inaccessible to JT5 (Model 761—630mm radius)

1263mm radius swept by JT5 ¢ measured from center of JT1 (Model 761—1511mm radius)

1388mm radius swept by mounting flange (Model 761—1636mm radius)

220°

Spec Sheet #14   Puma Series 700 specifications (courtesy of UNIMATION INCORPORATED, a Westinghouse Company)

■ Control Panel YASNAC RX

Controlled axes:	5 axes, possible to add 1 axis optionally
Position teaching method:	Teaching play-back, Geometric position data input
Positioning method:	Point-to-point (P.T.P.) and Linear/circular interpolation (C.P.)
Position memory:	IC memory with battery back-up
Programming capacity (for 5 axes):	2,200 positions, 1,200 instructions/or 5,000 positions, 2,500 instructions (Option in total)
Number of jobs:	249 max.
Drive units:	Transistor PWM control
Positioning method:	Incremental digital positioning
Accel. and decel. control:	Software type

Specifications — Motoman-L10W

Specifications		Motoman-L10W
Features		• Most suitable robot for arc welding. • Accurate tasks in wide working area.
Controlled Axes		5 degrees of freedom, jointed-arm type
Motion Range and Maximum Speed	Arm	S-axis turning: 300°, 90°/s
	Arm	L-axis lower arm movement: +45°, −40°, 1000 mm/s 39.4"/s
	Arm	U-axis upper arm movement: +20°, −45°, 1400 mm/s 55.1"/s
		—
	Wrist	B-axis swinging: 180°, 240°/s
	Wrist	T-axis twisting: 370°, 360°/s
Repetitive Positioning Accuracy		±0.2 mm ±0.008"
Payload		10 kg 22 lbs max
Weight		280 kg 616 lbs
Applications	Arc Welding	●
	Spot Welding	
	Cutting	●
	Laser Beam Cutting	●
	Gluing	●
	Material Handling	●
	Deburring, Finishing	
	Assembling	
	Inspection	●

Speed setting:	Teaching:	Up to 8 steps (20% of play-back speed)
	Play-back:	Setting by traverse time in 0.01 sec. unit; Speed setting in 1cm/min. unit
Coordinate system:		Rectangular/Cylindrical/Jointed arm coordinates
Interpolation function:		Linear/Circular interpolation
Tool center point control:		Possible
Forward/reverse of step:		Possible by teach box, Jog return possible
Adding, deleting and correcting teaching points:		Possible by teach box
Dry run:		Possible
Machine lock:		Possible
Waiting position teaching:		Possible
Software weaving:		Superposed on straight or circular line
Display information	Display unit:	9" CRT green-colored character display
	Characters:	Alphabets and numerals
	No. of characters:	32 characters × 16 lines
	No. of display:	8 categories, more than 128 displays in total
Instruction function:	M codes:	Linear/circular interpolation; Acceleration and deceleration
	A codes:	Timer: 0.1 ~25.5 sec. (0.1 sec. unit); 0.01 ~ 2.55 sec. (0.01 sec. unit) Input signal, setting/resetting/scanning Internal relay, setting/resetting/scanning
	H codes:	Counters/Robot halt/Index for jump function
	J codes:	Jump/Call/Skip jump functions
	F codes:	Welder control/Weaving function/Arc sensor
Files:	Welding condition:	Possible to program up to 16 files
	Weaving condition:	Possible to program up to 16 files
Parameters:		100 kinds
Search function:		Program, job, step, position, instruction
Job copy:		Possible
Input signals:		48 for general use/16 for specified use
Output signals:		24 for general use/7 for specified use
Internal relays:		64, built-in
Analog output channels:		2 channels (0V to +/− 14V/DC), 2 channels for option
Changing analog output voltage during play-back:		Possible for 2 channels simultaneously (welding voltage/current)
Cassette interface:		for TCM-5000
Printer interface:		for EPSON RP-80
Computer or sensor interface:		RS-232 C
Alarm indication:		48 kinds
Error indication:		120 kinds
Ambient temperature:		0 ~ 45°C (32 ~ 113°F)
Construction:		Free-stand/completely closed-type
External dimensions:	Box:	700 (W) × 1100 (H) × 580 (D) mm (27.56" × 43.31" × 22.83")
	Panel:	550 (W) × 400 (H) × 250 (D) mm (21.65" × 15.75" × 9.84")
Weight:		200 kg (440 lbs)
Finish color:		Muncell 7.5 BG 6/1.5 (Blue Green)
Power supply:		AC 200/220/230V +10%, −15%; 50/60Hz ±1Hz, 3-phase 3KVA

Spec Sheet #15   Motoman L10W specifications (courtesy of YASKAWA ELECTRIC AMERICA, INC.)

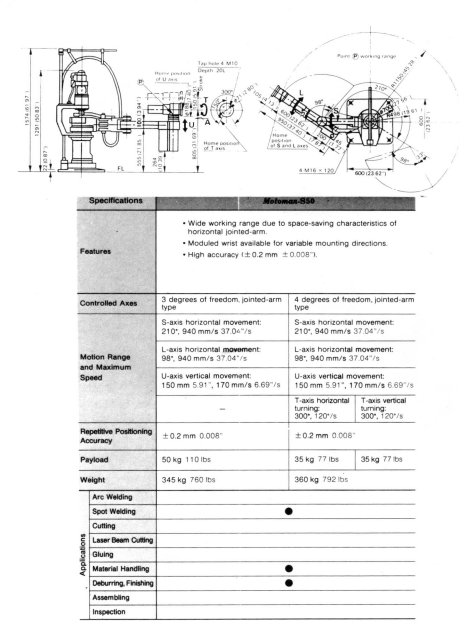

Specifications	Motoman-S50	
Features	• Wide working range due to space-saving characteristics of horizontal jointed-arm. • Moduled wrist available for variable mounting directions. • High accuracy (±0.2 mm ±0.008").	
Controlled Axes	3 degrees of freedom, jointed-arm type	4 degrees of freedom, jointed-arm type
Motion Range and Maximum Speed	S-axis horizontal movement: 210°, 940 mm/s 37.04"/s	S-axis horizontal movement: 210°, 940 mm/s 37.04"/s
	L-axis horizontal movement: 98°, 940 mm/s 37.04"/s	L-axis horizontal movement: 98°, 940 mm/s 37.04"/s
	U-axis vertical movement: 150 mm 5.91", 170 mm/s 6.69"/s	U-axis vertical movement: 150 mm 5.91", 170 mm/s 6.69"/s
	—	T-axis horizontal turning: 300°, 120°/s
		T-axis vertical turning: 300°, 120°/s
Repetitive Positioning Accuracy	±0.2 mm 0.008"	±0.2 mm 0.008"
Payload	50 kg 110 lbs	35 kg 77 lbs
		35 kg 77 lbs
Weight	345 kg 760 lbs	360 kg 792 lbs

Applications		
Arc Welding		
Spot Welding	●	
Cutting		
Laser Beam Cutting		
Gluing		
Material Handling	●	
Deburring, Finishing	●	
Assembling		
Inspection		

Spec Sheet #16 Motoman S50 specifications (courtesy of YASKAWA ELECTRIC AMERICA, INC.)

Controlled Axes		6 axes, possible to add 2 axes optionally
Position Teaching Method		Teaching play-back, Geometric position data input
Interpolation		Point-to-point (P.T.P.) and Linear/circular interpolation (C.P.)
Position Memory		IC memory with battery back-up
Programming Capacity		2,200 positions, 1,200 instructions. Up to 5,000 positions, 2,500 instructions available as an option.
Number of Jobs		249 max
Drive Units		Transistor PWM control
Positioning		Incremental digital positioning
Accel. and Decel. Control		Software type
Speed Setting	Teaching	Up to 8 steps (20 % of play-back speed)
	Play-back	Setting by traverse time in 0.01 s. unit; Speed setting in 1 cm/min.　　/min. unit.
Coordinate System		Rectangular/Cylindrical/Jointed arm coordinates
Tool Center Point Control		Programmable up to 8 files
Forward/Reverse of Step		Possible by teach box, Jog return possible
Adding, Deleting and Correcting Teaching Points		Possible by teach box
Dry run		Possible
Machine Lock		Possible
Waiting Position Teaching		Possible
Software Weaving		Superposed on straight or circular line
Display Information	Display Unit Characters No. of Characters No. of Display	9" CRT green-colored character display Alphabets and numerals 32 characters × 16 lines 8 categories, more than 128 displays in total
Instruction Function	M Codes	Linear/circular interpolation; Acceleration and deceleration
	A Codes	Timer: 0.1 to 25.5 s (0.1 s unit); 0.01 to 2.55 s (0.01 s unit). Input signal setting/resetting/scanning. Internal relay, setting/resetting/scanning
	H Codes	Counters/Robot halt/Index for jump function
	J Codes	Jump/Call/Skip jump functions
	F Codes	Welder control/Weaving function/Arc sensor
Files	Welding Condition	Programmable up to 16 files
	Weaving Condition	Programmable up to 16 files
Parameters		100 kinds
Search Function		Program, job, step, position, instruction
Job Copy		Possible
Input Signals		48 for general use/16 for specified use
Output Signals		24 for general use/7 for specified use
Internal Relays		64, built-in
Analog Output Channels		2 channels (0 V to ± 14 V/DC), 2 channels for option
Changing Analog Output Voltage During Play-back		Possible for 2 channels simultaneously (welding voltage/current)
Cassette Interface		for TCM-5000 EV
Printer Interface		for EPSON RP-80
Computer or Sensor Interface		RS-232C
Alarm Indication		48 kinds
Error Indication		120 kinds
Ambient Temperature		0 to 45℃ (32 to 113°F)
Construction		Free standing
Finish Color		Munsell 7.5 BG 6/1.5 (Blue Green)
Power Supply		200/220/230 VAC +10%, −15%; 50/60Hz ±1Hz, 3-phase 3kVA (V6, S50, L10W), 5kVA (V12, L106), 13kVA (L30, L100), 15kVA (L15), 20kVA (L60/L60W), 50kVA (L120)

Spec Sheet #17　Motoman S50 specifications (courtesy of YASKAWA ELECTRIC AMERICA, INC.)

Number of Axes: 6 (Additional Axes Optional)

Configuration: Jointed Arm

Mounting Position: Floor (Inverted Optional)

Coordinate System: Jointed Arm

Drive System: Hydraulic/Mechanical

Horizontal Stroke: PR16-28" PR24-37" PR36-52"

Horizontal Reach: PR16-46" PR24-62" PR36-87"

Vertical Stroke: PR16-48" PR24-66" PR36-91"

Vertical Reach: PR16-62" PR24-78" PR36-100"

Base Rotation: 100 Degrees

Wrist Roll: 450 Degrees

Wrist Movement: 120 Degree Cone

Control System: Closed loop digital servo
                 microcomputer based solid state

Memory Type: Semiconductor/disc memory option

Memory Size: 5 minutes continuous path
              (Expandable to over 5 hr.)
              point to point dependent on program

Number of Programs: 9 (Expandable to 500)

Random Selection: Yes

Operating Modes: Continuous path/Point to point

Programming: Lead through teach

Editing: Yes

Adjustable Playback Speed: Yes

Outputs: 8

Inputs: Remote Program Select Lines

Interface Hardware: Terminal strip for external wiring
                     RS232C Selectable baud rate
                     asynchronous RS232C synchronous

Ambient Conditions: 40 to 120 F-5% to 95%
                     humidity, non-condensing

Power Required: Maximum 10 kw

Thermwood Corporation reserves the right to make changes to the specifications at any time.

Spec Sheet #18   PR series robot specifications (courtesy of
THERMWOOD CORPORATION ROBOTICS DIVISION)

Technical Information

System Operating Specifications

Coordinate system	Cartesian, Joint, Region
Degrees of freedom	4 (All servoed)
Positioning control system	Point-to-point, circular, straight-line, via-point
Digital I/O ports (Configurable)	48 Standard, 192 Maximum
RS-232 Asynchronous Communications Port	4 Standard (1 for Pendant), 4 Optional
RS-422 Asynchronous Communications Port	1 Optional (Via Combination Adapter II)
Application program control	Operator Control Panel, Hand-Held Pendant, Host Communications
Point teaching method	Hand-Held Pendant
Maximum number of selectable programs	Seven via Operator Control Panel
Primary storage	512KB Random Access Memory, expandable to 640KB
Secondary storage	1.2MB Diskette (Standard), 1.2MB Diskette (Optional) 360KB Diskette (Optional), 20MB Fixed Disk (Optional)
Power requirements	IBM 7575 receives DC power from 7572 IBM 7572 - 220 - 240 Vac at 47 - 63 Hz IBM 7532/310 - 220 - 240 Vac at 47 - 63 Hz (Switch selectable to 100 - 130 Vac at 47 - 63 Hz)
System operating temperature	0°C - 40°C (32°F - 104°F)
System relative humidity	8% - 80% Non-condensing
Weight	IBM 7575 71 kg (156 lb)
	IBM 7572 48 kg (106 lb)
	IBM 7532/310 21.3 kg (47 lb)

All specifications are subject to change without notice. Performance is based on fixed operating conditions.

Manipulator Performance Specifications

X and Y axis repeatability	± 0.025 mm (± 0.001 in.)	
Maximum XY speed	5.1 m/sec (200 in./sec)	
Maximum Z speed	575 mm/sec (22.6 in./sec)	
Maximum roll speed	480 deg/sec	
Maximum payload (Reduced accelerations)	5 kg (11 lb)	
Symmetrical workspace:	Outside radius	550 mm (21.6 in.)
	Inside radius	222 mm (8.7 in.)
Theta 1 axis:	Arm length	325 mm (12.8 in.)
	Rotation	-120° - +120° (± 1°)
Theta 2 axis:	Arm length	225 mm (8.9 in.)
	Rotation	-137° - +137° (± 1°)
Z axis:	Stroke	150 mm (5.9 in.)
Roll axis:	Rotation	+3600° - -3600° (± 2°) (± 10 revolutions)

Spec Sheet #19 IBM 7575 manufacturing system specifications, features, and components (courtesy of INTERNATIONAL BUSINESS MACHINES CORPORATION)

Workspace

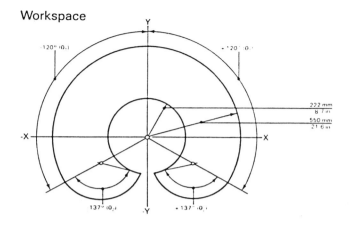

Optional Features

- 48-Point DI/DO Card (#6100)
- 48-Point DI/DO Cable (#6101)
- 4-Port RS-232 Asynchronous Communications Adapter (#4763)
- 4-Port RS-232 Asynchronous Communications Adapter Cable (#4704)
- 4-Port RS-232 Asynchronous Communications Adapter Rollup Cable (#6104)
- Hand-Held Pendant (#6105)
- 2-Meter Operator Panel Extension Kit (#6106)
- 3-Meter Manipulator Cable Set (#6108)
- 10-Meter Manipulator Cable Set (#6109)

- Color/Graphics Monitor Adapter (#4910)
- Serial/Parallel Adapter (#0215)
- Serial Adapter Cable (#0217)
- Serial Adapter Connector (#0242)
- Combination Adapter II (#6020)
- 128KB Memory Expansion for Combination Adapter II (#6040)
- Combination Adapter II Cable (#6001)
- 128KB Memory Expansion (#0209)
- 20MB Fixed Disk Drive (Customer-installed) (#6019)
- Second 1.2MB High-Capacity Diskette Drive (#6038)
- 360KB Dual-Sided Diskette Drive (Second drive) (#6039)

Other Supported Components

- IBM 5532 Industrial Color Display
- IBM 5533 Industrial Graphics Printer
- IBM 7534 Industrial Graphics Display

Spec Sheet #19   (Continued)

Technical Information

System Operating Specifications

Coordinate system	Cartesian, Joint, Region
Degrees of freedom	4 (All servoed)
Positioning control system	Point-to-point, circular, straight-line, via-point
Digital I/O ports (Configurable)	48 Standard, 192 Maximum
RS-232 Asynchronous Communications Port	4 Standard (1 for Pendant), 4 Optional
RS-422 Asynchronous Communications Port	1 Optional (Via Combination Adapter II)
Application program control	Operator Control Panel, Hand-Held Pendant, Host Communications
Point teaching method	Hand-Held Pendant
Maximum number of selectable programs	Seven via Operator Control Panel
Primary storage	512KB Random Access Memory, expandable to 640KB
Secondary storage	1.2MB Diskette (Standard), 1.2MB Diskette (Optional) 360KB Diskette (Optional), 20MB Fixed Disk (Optional)
Power requirements	IBM 7576 receives DC power from 7572 IBM 7572 - 220 - 240 Vac at 47-63 Hz IBM 7532/310 - 220 - 240 Vac at 47 - 63 Hz (Switch selectable to 100 - 130 Vac at 47 -63 Hz)
System operating temperature	0°C - 40°C (32°F - 104°F)
System relative humidity	8% - 80% Non-condensing
Weight	IBM 7576 76 kg (167 lb)
	IBM 7572 48 kg (106 lb)
	IBM 7532/310 21.3 kg (47 lb)

All specifications are subject to change without notice. Performance is based on fixed operating conditions.

Manipulator Performance Specifications

X and Y axis repeatability	± 0.050 mm (± 0.002 in.)	
Maximum XY speed	4.4 m/sec (173 in./sec)	
Maximum Z speed	575 mm/sec (22.6 in./sec)	
Maximum roll speed	480 deg/sec	
Maximum payload (Reduced accelerations)	10 kg (22 lb)	
Symmetrical workspace:	Outside radius	800 mm (31.5 in.)
	Inside radius	300 mm (11.8 in.)
Theta 1 axis:	Arm length	400 mm (15.7 in.)
	Rotation	-120° - +120° (± 1°)
Theta 2 axis:	Arm length	400 mm (15.7 in.)
	Rotation	-136° - +136° (± 1°)
Z axis:	Stroke	250 mm (9.8 in.)
Roll axis:	Rotation	+3600° - -3600° (± 2°) (± 10 revolutions)

Spec Sheet #20   IBM 7576 manufacturing system specifications, features, and components (courtesy of INTERNATIONAL BUSINESS MACHINES CORPORATION)

Workspace

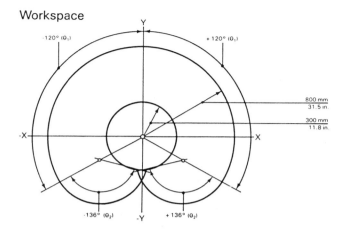

Optional Features

- 48-Point DI/DO Card (#6100)
- 48-Point DI/DO Cable (#6101)
- 4-Port RS-232 Asynchronous Communications Adapter (#4763)
- 4-Port RS-232 Asynchronous Communications Adapter Cable (#4704)
- 4-Port RS-232 Asynchronous Communications Adapter Rollup Cable (#6104)
- Hand-Held Pendant (#6105)
- 2-Meter Operator Panel Extension Kit (#6106)
- 3-Meter Manipulator Cable Set (#6108)
- 10-Meter Manipulator Cable Set (#6109)
- Color/Graphics Monitor Adapter (#4910)
- Serial/Parallel Adapter (#0215)
- Serial Adapter Cable (#0217)
- Serial Adapter Connector (#0242)
- Combination Adapter II (#6020)
- 128KB Memory Expansion for Combination Adapter II (#6040)
- Combination Adapter II Cable (#6001)
- 128KB Memory Expansion (#0209)
- 20MB Fixed Disk Drive (Customer-installed) (#6019)
- Second 1.2MB High-Capacity Diskette Drive (#6038)
- 360KB Dual-Sided Diskette Drive (Second drive) (#6039)

Other Supported Components

- IBM 5532 Industrial Color Display
- IBM 5533 Industrial Graphics Printer
- IBM 7534 Industrial Graphics Display

Spec Sheet #20   (Continued)

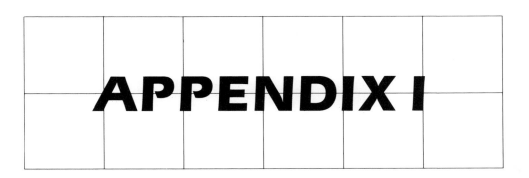

APPENDIX I

☐ CROSS-COMPARISON OF ROBOTS

This appendix table includes data on 166 robot models from 33 robot manufacturers. Note that the information is directional only and that some values and descriptions may vary between report sources and specifications manuals.

Table reproduced courtesy of Patrick J. Niedbala, Schoolcraft College, Livonia, Michigan.

| Manufacturer | Model | Payload | | Axes | | Powered |
		lbs.	Kgs.	No.	Action	By
Accuratio Systems Clawson, MI Office (313) 288-5070	System I* Gantry	75	35	5	Rectangular coordinate	Elec.
Advanced Robotics Corporation Columbus, Ohio (614)870-7778	Cyro 750	100	45	5	Rectangular coordinate	Elec.
Advanced Robotics Corporation	Cyro 1000	22	10	5-6	Jointed-arm spherical coordinate	Elec.
Advanced Robotics Corporation	Cyro 1350	250	113	6	Rectangular coordinate	Elec.
Advanced Robotics Corporation	Cyro 2000	350	159	5	Rectangular coordinate	Elec.
American Robot Corporation Farmington Hill, MI Office (313) 477-3200	MR6200*	20	9	6	Jointed-arm spherical coordinate	Elec.
American Robot	MR6500	50	23	6	Jointed-arm spherical coordinate	Elec.
American Robot	MR6260	22	11	6	Jointed-arm spherical coordinate	Elec.
American Robot	MR6260 Overhead	22	11	6	Jointed-arm spherical coordinate	Elec.
American Robot	MR6500 Overhead	50	23	6	Jointed-arm spherical coordinate	Elec.
ASEA (Sweden) Troy, MI office (313)528-3630	ASEA* IRb 6/2	13.2	6	5	Jointed-arm spherical coordinate	Elec.

Control Systems	SERVO	NONSERVO	Sequence of Motions Taught	Programming Method	Major Applications	Repeatable Placement Accuracy
CNC micro-computer	x		Continous path	Remote pendant or off-line	Mchn load/unload, deburr, sealant, adhesive, waterjet	+/- .008"
Mini-computer w/micro-processor	x		Continous path	Remote pendant or off-line	Welding	+/- .008"
Mini-computer w/micro-processor	x		Point-to-point and continous path	Remote pendant or off-line	Arc welding	+/- .008"
Mini-computer w/micro-processor	x		Point-to-point and continous path	Remote pendant or off-line	Arc welding	+/- .008"
Mini-computer w/micro-processor	x		Continous path	Remote pendant or off-line	Welding	+/- .016"
Micro-computer w/8 micro-processors	x		Point to point continous path circular + linear inter-polation	Remote pendant or off-line	Arc welding, mchn load/unload, assembly, palletizing	+/- .001"
Micro-computer w/8 micro-processors	x		Point to point continous path circular + linear inter-polation	Remote pendant or off-line	Arc welding, mchn load/unload, assembly, palletizing	+/- .001"
Micro-computer w/8 micro-processors	x		Point to point continous path circular + linear inter-polation	Remote pendant or off-line	Arc welding, mchn load/unload, assembly, palletizing	+/- .004"
Micro-computer w/8 micro-processors	X		Point to point continuous path circular + linear inter-polation	Remote pendant or off-line	Arc welding, mchn load/unload, assembly palletizing	+/- .004"
Micro-computer w/8 micro-processors	x		Point to point continous path circular + linear inter-polation	Remote pendant or off-line	Arc welding, mchn load/unload, assembly, palletizing	+/- .001"
CNC micro-computer	x		Point-to-point	Remote pendant	Mchn load/unload, pick & place, spot welding	+/- .008"

continues

Manufacturer	Model	Payload		Axes		Powered By
		lbs.	Kgs.	No.	Action	
ASEA	ASEA* IRb 60/2	132	60	5	Jointed-arm spherical coordinate	Elec.
ASEA	ASEA* IRb 90S/2	198	90	5-6	Jointed-arm spherical coordinate	Elec.
ASEA	MHU Minor (Electrolux)	2.2	1	3	Cylindrical coordinate	Pnem.
ASEA	MHU Junior (Electrolux)	11	5	3	Cylindrical coordinate	Pnem.
ASEA	MHU Senior (Electrolux)	33	15	3	Cylindrical coordinate	Pnem.
Automatix Inc. Farmington Hills, MI Office (313)553-0044	AID 600	66	30	5	Rectilinear coordinate	Elec.
Automatix Inc.	AID 800	22	10	5	Jointed-arm spherical coordinate	Elec.
Automatix Inc	AID 900	33-65	15-30	6-5	Jointed-arm spherical coordinate	Elec
C. Itoh & Company Farmington, MI Office (313)473-7400	NACHI UM 5600 SP	11	5	6	Jointed-arm spherical coordinate	Hydraulic
C. Itoh & Company	NACHI UM 5601 AF1	11	5	6	Jointed-arm spherical coordinate	Hydraulic
C. Itoh & Company	NACHI UM 7500 AE (Cyro 820)	20	9	5	Jointed-arm spherical coordinate	Elec.
C. Itoh & Company	NACHI UN 8600 AK	110	50	6	Jointed-arm spherical coordinate	Elec.
Cincinnati Milacron Southfield, MI ofc (313) 557-2700	T3-566*	100	45	6	Jointed-arm spherical coordinate	Hydraulic
Cincinnati Milacron	T3-576* Extended reach	135	61	6	Jointed-arm spherical coordinate	Hydraulic

Control Systems	SERVO	NON SERVO	Sequence of Motions Taught	Pro-gramming Method	Major Applications	Repeatable Placement Accuracy
CNC micro-computer	x		Point-to-point	Remote pendant	Mchn load/unload, pick & place, spot welding	+/- .016"
CNC micro-computer	x		Point-to-point and supple-mental program	Remote pendant	Spot welding	+/- .004"
Micro-processor		x	Point-to-point	Off-line	Pick and place	+/- .004"
Micro-processor		x	Point-to-point	Remote pendant	Spot welding, deburr	+/- .004"
Micro-processor		x	Point-to-point	Manual stop set push-button sequence	Pick and place, mchn load/unload	+/- .004"
Micro-computer	X		Point-to-point continuous	Remote pendant or off-line	Assembly & advance mtrl handling	+/- .003"
Micro-computer	x		Point-to-point continuous	Remote pendant or off-line	Arc welding & advance mtrl handling	+/- .008"
Micro-computer	x		Point-to-point continuous	Remote or pendant or off-line	Arc welding & advance mtrl handling	+/- .008"
Micro-processor floppy disc	x		Continuous path	Manual teach playback	Spray painting	+/- .080"
Micro-processor plate wire memory	x		Point-to-point	Remote pendant or manual teach playback	Spray Painting	+/- .080"
Micro-processor plate wire & cassette deck	x		Point-to-point	Remote pendant	Arc welding	+/- .010"
Micro-processor plate wire & cassette deck	x		Point-to-point	Remote pendant	Spot welding & mtrl handling	+/- .040"
Mini-computer	x		Point-to-point	Remote pendant	Mchn Load/unload, pick & place, spot/arc welding	+/- .050"
Mini-computer	x		Point-to-point	Remote pendant	Mchn load/unload, pick & place, spot/arc welding	+/- .050"

continues

| Manufacturer | Model | Payload | | Axes | | Powered |
		lbs.	Kgs.	No.	Action	By
Cincinnati Milacron	T3-586*	225	102	6	Jointed-arm spherical coordinate	Hydraulic
Cincinnati Milacron	T3-595*	450	204	5	Jointed-arm spherical coordinate	Hydraulic
Cincinnati Milacron	T3-726*	14	6	6	Jointed-arm spherical coordinate	Elec.
Cincinnati Milacron	T3-746*	70	31.7	6	Jointed-arm spherical coordinate	Elec.
Cincinnati Milacron	T3-776*	150	68	6	Jointed-arm spherical coordinate	Elec.
Cincinnati Milacron	T3-786*	200	90	6	Joint-arm spherical coordinate	Elec.
Cincinnati Milacron	T3-363	110	50	3	Cylindrical coordinate	Elec.
Cincinnati Milacron	T3-364	75	34	4	Cylindrical coordinate	Elec.
Cincinnati Milacron	T3-876* Gantry	150	68	6	Rectilinear spherical coordinate	Elec.
Cincinnati Milacron	T3-886* Gantry	200	90	6	Rectilinear spherical coordinate	Elec.
Cincinnati Milacron	T3-895* Gantry	440	200	5	Rectilinear spherical coordinate	Elec.
COMAU Industriale (Italy) Troy, MI office (313)583-9900	Polar 6000	130	59	6	Spherical coordinate	Hydraulic
COMAU Industriale	Smart 6.50	132	60	6	Jointed-arm spherical coordinate	Elec.
Control Automation, Inc., Princeton, NJ (609)799-6026	Mini-sembler	5	2.3	4	Rectilinear coordinate	Elec.

Control Systems	SERVO	NON SERVO	Sequence of Motions Taught	Programming Method	Major Applications	Repeatable Placement Accuracy
Mini-computer	x		Point-to-point	Remote pendant	Mchn load/ un-load, pick & place, spot/ arc welding	+/- .050"
Mini-computer	x		Point-to-point	Remote pendant	Mchn load/ un-load, pick & place, spot/ arc welding	+/- .050"
Micro-processor	x		Point-to-point	Remote pendant	Mtrl handling, assembly, arc welding	+/- .004"
Micro-processor	x		Point-to-point	Remote pendant	Mtrl handling, arc welding (variable)	+/- .010"
Micro-processor	x		Point-to-point	Remote pendant	Mtrl handling, arc welding (variable)	+/- .010"
Micro-Processor	x		Point-to-point	Remote pendant	Mtrl handling, spot/arc weld-ing (variable)	+/- .010"
Micro-processor	x		Point-to-point	Remote pendant	Mchn load/un-load, pick & place	+/- .040"
Micro-processor	x		point-to-	Remote pendant	Mchn load/un-load, pick & place	+/- .040"
Micro-processor	x		Point-to-point	Remote pendant	Mtrl handling, spot/arc welding	+/- .010"
Micro-processor	x		Point-to-point-	Remote pendant	Mtrl handling, spot/ arc weld-ing	+/- .010"
Micro-processor	x		Point-to-point-	Remote pendant	Mtrl handling, spot/ arc weld-ing	+/- .010"
Program-mable controller	x		Point-to-point-	Off-line	Spot welding, mchn load/un-load	+/- .040"
Micro-computer	x		Point-to-point- and contin-uous path	Remote pendant or off-line	Spot welding, assembly, mtrl handling	+/- .015"
Mini-computer	x		Continuous path	Remote pendant or off-line	Electronic assembly, mtrl handing	+/- .001"

continues

| Manufacturer | Model | Payload | | Axes | | Powered |
		lbs.	Kgs.	No.	Action	By
Cybotech Novi, MI office (313)348-3335	Cybotech H80	175	80	6	Jointed-arm cylindrical coordinate	Elec.
Cybotech	Cybotech V15	33	15	6	Cylindrical coordinate	Elec.
Cybotech	Cybotech TH8	17	8	6	Cylindrical coordinate	Elec.
Cybotech	Cybotech TP15	11	4.4	6	Jointed-arm spherical coordinate	Elec.
Cybotech	Cybotech WV15	22	10	6	Jointed-arm spherical coordinate	Elec.
Cybotech	Cybotech WCX90	11	5	8	Cartesian coordinate	Elec.
Cybotech	Cybotech G80	175	80	6	Cartesian coordinate	Elec.
Durr Industries, Inc. Plymouth, MI office (313)459-6800	P-100* gantry	220	100	3-5	Cartesian coordinate	Elec.
Durr Industries, Inc.	CompArm (Hall Automation- England)	7	3	6	Spherical coordinate	Hydraulic
Gametics Roseville, MI (313)778-7220	Model 524*	50	23	3-5	Cylindrical coordinate	Hydraulic
Gametics	Model 536*	200	114	3-5	Cylindrical coordinate	Hydraulic
Gametics	Model 524E	100	57	5	Cylindrical coordinate	Elec.

Control Systems	SERVO	NON SERVO	Sequence of Motions Taught	Programming Method	Major Applications	Repeatable Placement Accuracy
Micro Processor based R6 control	x		Continuous path or point-to-point-	Manual teach playback or remote pendant	Mtrl handling, spot/ arc weld welding	+/- .008"
Micro-processor based RC6 controller	x		Point-to-point-	Remote pendant	Assembly, inspection/handling, arc welding	+/- .004"
Micro-processor based RC6 controller	x		Point-to-point-	Remote pendant	Arc welding, mtrl handling, assembly	+/- .008"
Micro-processor based RC6 controller	x		Continuous path or point-to-point	Remote pendant	Painting, sealing, coating	+/- .008"
Micro-processor based RC6 controller	x		Continuous path or point-to-point	Remote pendant	Arc welding	+/- .008"
Micro-processor based RC6 controller	x		Continuous path or point-to-point	Remote pendant	Arc welding	+/- .008"
Micro-processor based RC6 controller	x		Continuous path or point-to-point	Remote pendant	Arc welding	+/- .008"
Micro-processor & PC	x		Point-to-point CP by inter-polation	Remote pendant or off-line	Mchn load, palletizing, assembly, water cutting	+/- .008"
Micro-processor	x		Continuous path	Manual teach playback	Spray painting	+/- .040"
Micro-Processor	x		Point-to-point	Remote pendant	Mchn load/ unload	+/- .030"
Micro Processor	x		Point-to-point	Remote pendant	Mchn load/ unload	+/- .050"
Micro processor	x		Point-to-point	Remote pendant	Mchn load/ unload	+/- .030"

continues

| Manufacturer | Model | Payload | | Axes | | Powered |
		lbs.	Kgs.	No.	Action	By
GCA Corporation Troy, MI office (313) 362-4144	DKP 200N	4.4	2	4	Horizontal joint-arm cylindrical coordinate	Elec.
GCA Corporation	DKP 200V	4.4	2	5	Jointed-arm spherical coordinate	Elec.
GCA Corporation	DKP 300H	11	5	4	Horizontal joint-arm coordinate	Elec.
GCA Corporation	DKP 300V	11	5	5	Jointed-arm spherical	Elec.
GCA Corporation	DKP 550	15	7	5	Jointed-arm cylindrical coordinate	Elec.
GCA Corporation	DKP 600	26	11.8	5	Jointed-arm spherical coordinate	Elec.
GCA Corporation	DKP 800	66	30	3	Jointed-arm spherical coordinate	Elec.
GCA Corporation	DKP 1000	66	30	6	Jointed-arm spherical coordinate	Elec.
GCA Corporation	DKP 1200	110	50	6	Jointed-arm spherical coordinate	Elec.
GCA Corporation	DKB 1440*	110	50	4-6	Cylindrical & cartesian coordinate	Elec.
GCA Corporation	DKB 3200	220	100	4-6	Cylindrical & cartesian coordinate	Elec.

Control Systems	SERVO	NON SERVO	Sequence of Motions Taught	Programming Method	Major Applications	Repeatable Placement Accuracy
Micro-processor two models	x		Continous path or point-to-point	Remote pendant or off-line	Assembly, pkging, mchn load, mtrl handling	+/- .002"
Micro-processor two models	x		Continous path or point-to-point	Remote pendant or off-line	Assembly, pkging, mchn load,, mtrl handling	+/- .002"
Micro-processor	x		Continous path or point-to-	Remote pendant line	Assembly, pkging, mchn load, mtrl handling	+/- .004"
Micro-processor	x		Point-to-point & continuous	Remote pendant line	Assembly, pkging mchn load, mtrl handling	+/- .004"
Micro-processor two models	x		Point-to-point & continuous path	Remote pendant or off-line	Assembly arc welding, sealing, mtrl handling	+/- .008"
Micro-processor two models	x		Point-to-point & continuous path	Remote pendant or off-line	Arc welding, assembly	+/- .004"
Micro-processor two models	x		Point-to-point & continuous path	Remote pendant or off-line	Arc/spot welding, mchn load, mtrl handling	+/- .020"
Micro-processor two models	x		Point-to-point & continuous path	Remote pendant or off-line	Assembly, arc/spot welding, mtrl handling	+/- .020"
Micro-processor two models	x	x	Point-to-point & continuous path	Remote pendant or off-line	Assembly, arc/spot welding mtrl handling	+/- .002"
Micro-processor two models	x		Point-to-point & continuous path	Remote pendant or off-line	Mtrl handling, part transfer, palletizing	+/- .02"
Micro-processor	x		Point-to-point & continuous path	Remote pendant or off-line	Mtrl handling, part transfer, palletizing	+/- .04"

continues

| Manufacturer | Model | Payload | | Axes | | Powered |
		lbs.	Kgs.	No.	Action	By
GCA Corporation	XR 50* series	300	136	3-6	Rectilinear coordinate	Elec.
GCA Corporation	XR 100*	2200	1000	3-6	Rectilinear coordinate	Elec.
General Electric Troy, MI (313) 362-5550	P-50	22	10	5	Jointed-arm spherical coordinate	Elec.
General Electric	GP-110	110	50	6	Jointed-arm spherical coordinate	Elec.
General Electric	GP-155	155	70	6	Jointed-arm spherical coordinate	Elec.
General Electric	GP-220	220	100	6	Jointed-arm spherical coordinate	Elec.
General Electric	A-3	4.4	2	3	Scara type	Elec.
General Electric	A-4	4.4	4	3	Scara type	Elec.
General Electric	A-40	22	10	4	Scara type	Elec.
General Electric	Allegro A-12 (DEA)	14	6.5	5 per arm	Rectilinear coordinate	Elec.
GMF Robotics Troy, MI office (313) 641-4223	A-0	22	10	3-5	Cylindrical coordinate	Elec.

Control Systems	SERVO	NONSERVO	Sequence of Motions Taught	Programming Method	Major Applications	Repeatable Placement Accuracy
Micro-processor CIMROC	x		Point-to-point & continuous path	Remote pendant or off-line	Mtrl handling, part transfer, palletizing	+/- .004"
Micro-processor CIMROC	x		Point-to-point & continuous path	Remote pendant or off-line	Heavy duty handling, arc/ spot welding	+/- .008"
Micro-processor	x		Continuous path	Remote pendant	Arc welding, sealant, adhesive, grinding process	+/- .008"
Micro processor	x		Continuous path	Remote pendant	Spot welding & mtrl handling	+/- .02"
Micro processor	x		Continuous path	Remote pendant	Spot welding & mtrl handling	+/- .02"
Micro processor	x		Continuous path	Remote pendant	Spot welding & mtrl handling	+/- .04"
Micro processor	x		Point-to-point or continuous path	Remote pendant	Assembly, component insertion, mtrl handling, sealant/ adhesive	+/- .002"
Micro processor	x		Point-to-point or continuous path	Remote pendant	Assembly, component insertion, mtrl handling, sealant/ adhesive	+/- .002"
Micro processor	x		Point-to-point or continuous path	Remote pendant	Assembly, component insertion, mtrl handling, sealant/ adhesive	+/- .004"
Micro processor	x		Point-to-point	Remote pendant or off-line	Precision assembly, pick & place	+/- .001"
Micro processor	x		Point-to-point	Remote pendant	Mtrl handling, assembly, deburr	+/- .002"

continues

| Manufacturer | Model | Payload | | Axes | | Powered By |
		lbs.	Kgs.	No.	Action	
GMF Robotics	A-1	66	30	3-5	Cylindrical coordinate	Elec.
GMF Robotics	S-108R	17	8	5	Jointed-arm spherical coordinate	Elec.
GMF Robotics	S-108L (L-1000) (L-2000)	17	8	5	Jointed-arm spherical coordinate	Elec.
GMF Robotics	S360R	132	60	6	Jointed-arm spherical coordinate	Elec.
GMF Robotics	S-360L (L-1000) (L-2000)	132	60	6	Jointed-arm spherical coordinate	Elec.
GMF Robotics	M-00	44	20	5	Cylindrical coordinate	Elec. pneumatic
GMF Robotics	M-0	44	20	4	Cylindrical coordinate	Elec.
GMF Robotics	M-1	103	47	3-5	Cylindrical coordinate	Elec.
GMF Robotics	M-1A	103	47	3-5	Cylindrical coordinate	Elec.
GMF Robotics	M-2	132	60	4	Cylindrical coordinate	Elec.
GMF Robotics	M-3	264	120	3-5	Cylindrical coordinate	Elec.
GMF Robotics	S-110R	22	10	5-6	Jointed-arm spherical coordinate	Elec.
GMF Robotics	A-00	22	10	3-4	Cylindrical coordinate	Elec.

Control Systems	S E R V O	N O N S E R V O	Sequence of Motions Taught	Programming Method	Major Applications	Repeatable Placement Accuracy
Micro processor	x		Point-to-point	Remote pendant	Mtrl handling, assembly	+/- .002"
Microprocessor	x		Point-to-point	Remote pendant	Arc welding, mtrl handling, sealing	+/- .008"
Microprocessor	x		Point-to-point	Remote pendant	Arc welding, mtrl handling, sealing	+/- .008"
Micro processor	x		Point-to-point	Remote pendant	Spot welding, mtrl handling, sealing	+/- .02"
Microprocessor	x		Point-to-point	Remote pendant	Spot welding, mtrl handling sealing	+/- .02"
Microprocessor	x		Point-to-point	Remote pendant	Mchn load/ unload	+/- .012"
Microprocessor	x		Point-to-point	Remote pendant	Mchn load/ unload	+/- .02"
Microprocessor	x		Point-to-point	Remote pendant	Mtrl handling, assembly	+/- .039"
Microprocessor	x		Point-to-point	Remote pendant	Mtrl handling, assembly	+/- .039"
Microprocessor	x		Point-to-point	Remote pendant	Mchn load/ unload, mtrl handling	+/- .039"
Microprocessor	x		Point-to-point	Remote pendant	Mtrl handling mchn load/ unload	+/- .039"
Microprocessor	x		Point-to-point	Remote pendant	Arc welding, mtrl handling & sealing	+/- .008"
Microprocessor	x		Point-to-point	Remote pendant	Assembly mtrl handling	+/- .002"

continues

| Manufacturer | Model | Payload | | No. | Axes | Powered |
		lbs.	Kgs.	No.	Action	By
GMF Robotics	S380R	180	81	6	Jointed-arm spherical coordinate	Elec.
GMF Robotics	S380L (L-2000) (L-3000)	180	81	6	Jointed-arm spherical coordinate	Elec.
GMF Robotics	N/C	-	-	7	Rectangular coordinate	Hydraulic
Graco Robotics, Inc. Livonia, MI office (313) 523-6300	O.M. 5000	16	7.2	6	Jointed-arm spherical coordinate	Hydraulic
Graco Robotics, Inc.	O.M. 5	16	7.2	5	Jointed-arm spherical coordinate	Hydraulic
Graco Robotics, Inc.	O.M. 50	16	7.2	5	Jointed-arm spherical coordinate	Hydraulic
Graco Robotics, Inc.	O.M. 500	16	7.2	6	Jointed-arm spherical coordinate	Hydraulic
IBM Southfield, MI office (313) 827-7707	7535*	13.2	6	4	Horizontal jointed-arm cylindrical coordinate	Elec. air
IBM	7545*	22	10	4	Horizontal jointed-arm cylindrical	Elec.
IBM	7540*	55	25	4	Horizontal jointed-arm cylindrical coordinate	Elec. air

Control Systems	S E R V O	N O N S E R V O	Sequence of Motions Taught	Pro-gramming Method	Major Applications	Repeatable Placement Accuracy
Micro-processor	x		Point-to-point	Remote pendant	Spot welding, mtrl handling, sealing	+/- .02"
Micro-processor	x		Point-to-point	Remote pendant	Spot welding, mtrl handling, sealing	+/- .02"
Micro-processor	x		Point-to-point	Remote pendant	Painting, inspection	+/- .25"
Micro-processor	x		Continuous path and point-to-point	Manual teach play-back or remote pendant	Spray coating	+/- .125
Micro-processor	x		Continuous path and point-to-point	Manual teach play-back or remote pendant	Spray coating	+/- .125
Micro-processor	x		Continuous path and point-to-point	Manual teach play-back or remote pendant	Spray coating	+/- .125
Micro-processor	x		Continuous path and point-to-point	Manual teach play-back or remote pendant	Spray coating	+/- .125
Micro-processor	2 axes	2 axes	Point-to-point	On-line with IBM PC and off-line	Small assembly and processing	+/- .002"
Micro processor	4 axes		Point-to-point	On line with IBM PC and off-line	Small parts assembly and processing	+/- .002"
Micro processor	2 axes	2 axes	Point-to-point	On line with IBM PC and off-line	Small parts assembly and processing	+/- .002"

continues

| Manufacturer | Model | Payload | | Axes | | Powered |
		lbs.	Kgs.	No.	Action	By
IBM	7547	55	25	4	Horizontal jointed-arm cylindrical coordinate	Elec. air
IBM	7565	5	2.2	3-6	Rectilinear spherical coordinate	Hydraulic
Intelledex Corvallis, OR (503) 758-4700	405	12	5.4	4	Horizontal jointed-arm cylindrical coordinate	Elec.
Intelledex	605 S & T	5	2.2	6	Horizontal jointed-arm cylindrical coordinate	Elec.
Intelledex	705 S	5	2.2	7	Horizontal jointed-arm cylindrical coordinate	Elec.
KUKA (Germany) Expert Automation Sterling Heights, MI (313) 977-0100	KUKA* IR662/100	220	100	6	Jointed-arm cylindrical coordinate	Elec.
KUKA	KUKA IR 200	132	60	5-7	Horizontal joint-arm cylindrical	Elec.
KUKA	KUKA* IR 160/60	132 (1100 welding force)	60 500	4-6	Jointed-arm cylindrical coordinate	Elec.
KUKA	IR 160/15*	33	15	4-6	Jointed-arm cylindrical coordinate	Elec.
NIKO (Germany) United Technologies Dearborn, MI (313) 593-9616	NIKO 25	5	2.5	5	Jointed-arm spherical coordinate	Elec.
NIKO	NIKO 50	10	5	5	Jointed-arm spherical coordinate	Elec.
NIKO	NIKO 100	44	20	6	Horizontal Jointed-arm spherical coordinate	Elec.
NIKO	NIKO 150	33	15	5	Jointed-arm spherical coordinate	Elec.

Control Systems	SERVO	NON SERVO	Sequence of Motions Taught	Pro- gramming Method	Major Applications	Repeatable Placement Accuracy
Micro processor	4 axes		Point-to-point	On line with IBM PC and off-line	Small parts assembly and processing	+/- .002"
IBM computer	3-6		Continuous path	Remote or off-line	Small parts assembly and processing	+/- .005"
Micro-processor		x	Point-to-point	Remote pendant or off-line	Assembly	+/- .001"
Micro-processor		x	Point-to-point	Remote pendant or off-line	Assembly	+/- .001"
Micro-processor		x	Point-to-point	Remote pendant or off-line	Assembly	+/- .001"
Micro-computer		x	Point-to-point and continuous path	Remote pendant or off-line	Spot welding	+/- .04"
Micro-computer		x	Point-to-point	Remote pendant	Spot welding, mchng & handling	+/- .04"
Micro-computer		x	Point-to-point and continuous path	Remote pendant or off-line	Spot welding	+/- .04"
Micro-computer		x	Point-to-point and continuous path	Remote pendant or off-line	Assembly, mtrl handling, arc welding	+/- .04"
Micro-processor		x	Point-to-point and continuous path	Remote pendant	Light assembly, seal-ant	+/- .01"
Micro-processor		x	Point-to-point and continuous path	Remote pendant	Light assembly, seal-ant	+/- .01"
Micro-processor		x	Point-to-point and continuous path	Remote pendant	Mtrl handling, sealant applic-ations	+/- .015"
Micro-processor		x	Point-to-point and continuous path	Remote pendant	Mtrl handling, assembly, seal-ant	+/- .01"

continues

| Manufacturer | Model | Payload | | Axes | | Powered |
		lbs.	kgs.	No.	Action	By
NIKO	NIKO 200	44	20	6	Cartesian coordinate	Elec.
NIKO	NIKO 450	132	60	6	Jointed-arm spherical coordinate	Elec. processor
NIKO	NIKO 451	176	80	6	Jointed-arm spherical coordinate	Elec.
NIKO	NIKO 500	176	80	6	Horizontal Jointed-arm spherical coordinate	Elec.
NIKO	NIKO 600*	176	80	6	Cartesian coordinated	Elec.
Nimak (West Germany)	Nike 600	130	59	6	Rectangular coordinate (Gantry)	Elec.
Prab Robots, Inc. Farmington Hills, MI Office (313)478-7722	Prab 4200/* 4200 HD	75 125	34 57	3-5	Spherical coordinate	Hydraulic
Prab Robots, Inc.	Prab 5800/* 5800 HD	50 100	23 46	3-5	Spherical coordinate	Hydraulic
Prab Robots, Inc.	Prab Series E	100	46	3-5	Cylindrical coordinate	Hydraulic
Prab Robots, Inc.	Prab* Series FA	250	114	3-7	Cylindrical coordinate	Hydraulic
Prab Robots, Inc.	Prab Series FB	600	272	3-7	Cylindrical coordinate	Hydraulic
Prab Robots, Inc.	Prab Series FC	2000	909	3-7	Cylindrical coordinate	Hydraulic

Control Systems	SERVO	NON SERVO	Sequence of Motions Taught	Programming Method	Major Applications	Repeatable Placement Accuracy
Micro-processor	x		Point-to-point and continuous path	Remote pendant	Light assembly, mtrl handling, arc welding, sealant	+/- .01"
Micro-	x		Point-to-point and continuous path	Remote pendant	Spot welding, mtrl handling	+/- .013"
Micro-processor	x		Point-to-point and continuous path	Remote pendant	Spot welding+ mtrl handling	/- .013"
Micro-processor	x		Point-to-point and continuous path	Remote pendant	Mtrl handling, sealant applic-ations, welding applications	+/- .015"
Micro-processor	x		Point-to-point and continuous path	Remote pendant	Spot welding, mtrl handling	+/- .013"
Mini-computer	x		Point-to-point	Remote pendant	Spot welding (will make a continuous path unit for arc welding)	+/- .02"
Pos. stops and memory drum		x	Point-to-point sequence	Set pos. stops and drum memory	Mchn load/ unload, pick & place	+/- .008"
Pos. stops and memory drum		x	Point-to-point	Set pos. stops and drum memory	Mchn load/ unload, pick and place	+/- .008"
Micro-processor	x		Point-to-point	Remote pendant	Mchn load/ unload, pick and place	+/- .03"
Micro-processor	x		Point-to-point	Remote pendant	Mchn load/ unload, pick and place	+/- .05"
Micro-processor	x		Point-to-point	Remote pendant	Mchn load/ unload, pick and place	+/- .05"
Micro-processor	x		Point-to-point	Remote pendant	Mchn load/ unload, pick and place	+/- .08"

continues

| Manufacturer | Model | Payload | | Axes | | Powered |
		lbs.	Kgs.	No.	Action	By
Prab Robots, Inc.	G05	30-65	15-30	5	Cartesian coordinate	Elec.
Prab Robots, Inc.	G06	30-65	15-30	6	Cartesian coordinate	Elec.
Prab Robots, Inc.	G24	55-130	25-60	4	Cartesian coordinate	Elec.
Prab Robots, Inc.	G26	55-130	25-60	6	Cartesian coordinate	Elec.
Prab Robots, Inc.	G34	140-220	65-100	4	Cartesian coordinate	Elec.
Prab Robots, Inc.	G36	140-220	65-100	6	Cartesian coordinate	Elec.
Reis (Germany) (Represented by: E&E Engineering (313)371-2000	Model 625*	88	40	6	Cylindrical coordinate	Elec.
Reis	Model 650*	110	50	6	Cylindrical coordinate	Elec.
Reis	Model V-15	33	15	6	Jointed-arm coordinate	Elec.
Reis	Model H-15	33	15	6	Cylindrical coordinate	Elec.

Control Systems	SERVO	NON SERVO	Sequence of Motions Taught	Pro- gramming Method	Major Applications	Repeatable Placement Accuracy
Micro- processor	x		Point-to-point or continuous path	Remote pendant or off-line	Arc welding, sealing, deburring, assembly, mtrl handling	+/- .008"
Micro- processor	x		Point-to-point or continuous path	Remote pendant or off-line	Arc welding, sealing, deburring, assembly, mtrl handling	+/- .008"
Micro- processor	x		Point-to-point or continuous path	Remote pendant or off-line	Mchn load, palletizing, welding, assembly, deburring	+/- .015"
Micro- processor	x		Point-to-point or continuous path	Remote pendant or off-line	Mchn load, palletizing, welding, assembly, deburring	+/- .015"
Micro- processor	x		Point-to-point or continuous path	Remote pendant or off-line	Spot welding, palletizing, mtrl handling, mtrl process- ing	+/- .025"
Micro- processor	x		Point-to-point or continuous path	Remote pendant or off-line	Spot welding, palletizing, mtrl handling, mtrl process- ing	+/- .025"
Micro- computer	x		Point-to-point	Remote pendant	Mchn load/ unload, grind- ing, stack- ing, palletiz- ing	+/- .015"
Micro- computer	x		Point-to-point	Remote pendant	Mchn load/ unload, grind- ing, stack- ing, palletiz- ing	+/- .015"
Micro- computer	x		Point-to-point and continuous path	Remote pendant	Assembly gluing	+/- .004"
Micro- computer	x		Point-to-point and continous path	Remote pendant	Mchn load/ unload, grind ing, stack- ing, palletiz- ing, assembly gluing	+/- .004"

continues

| Manufacturer | Model | Payload | | Axes | | Powered |
		Tbs.	Kgs.	No.	Action	By
Rimrock Corp. Columbus, Ohio (614)471-5926 (Was Auto Place, Copperweld)	CR-10	10	4.5	4	Cylindrical coordinate	Pneu. Z-80
Rimrock Corp.	CR-50	20	9	4	Cylindrical coordinate	Pneu.
Robotek (Fraser Automation) (313)979-4500	Model FR-200-3	150	68	3	Cylindrical coordinate	Elec.
Robotek	Model FR-200-4	66	30	4	Cylindrical coordinate	Elec.
Robotek	Model FR-300-3	66	30	4	Cylindrical coordinate	Elec.
Robotek	Model FR-300-4	66	30	3	Cylindrical coordinate	Elec.
Robotek	Model FR-400-3 to 5	35	1	3-5	Jointed-arm cylindrical coordinate	Elec.
Robotek	Model FR-500	35	16	3-5	Rectilinear coordinate	Elec.
Robotics, Inc Detroit, MI office	NC 54.12* Gantry	50	23	3-5	Rectilinear coordinate	Elec.
Schrader-Bellows Clawson, MI office (313)522-4411	Motion* Mate	5	2.5	5	Cylindrical coordinate	Air
Thermwood (Represented by: Roethel Engineering Farmington Hills, MI (313) 478-7950	Paintmiser	18	8	6	Jointed-arm spherical coordinate	Hydraulic
Thermwood	Taskhandler	25	12	5	Jointed-arm spherical coordinate	Hydraulic
Thermwood	Cartesian 5 Gantry System			5	Cartesian	Elec.

Control Systems	SERVO	NON SERVO	Sequence of Motions Taught	Pro- gramming Method	Major Applications	Repeatable Placement Accuracy
Proximity switch micro- processor		x	Point-to-point	Remote pendant	Mchn load/ unload, pick and place	+/- .003"
Proximity switch micro- processor Z-80		x	Point-to-point	Remote pendant	Mchn load/ unload, pick and place	+/- .003"
Micro- processor	x		Continuous path	Remote pendant or off-line	Pick and place	+/- .016"
Micro- processor	x		Continuous path	Remote pendant or off-line	Pick and place	+/- .016"
Micro- processor	x		Continuous path	Remote pendant or off-line	Pick and place	+/- .016"
Micro- processor	x		Continuous path	Remote pendant or off-line	Pick and place	+/- .016"
Micro- processor	x		Continuous path	Remote pendant or off-line	Pick and place	+/- .003"
Micro- processor	x		Continuous path	Remote pendant or off-line	Pick and place	+/- .003"
Micro- processor	x		Continuous path	Remote pendant	Dispensing adhesives or bonding	+/- .01"
Micro- processor		x	Point-to-point	Manual teach playback	Pick and place	+/- .005"
Micro- processor hard disc option	x		Continuous path	Lead through teach edit-box	Painting, sealing, coating	+/- .125"
Micro- processor hard disc option	x		Continous path	Lead through teach edit-box	Machine load, palletizing, pick & place	+/- .125"
Micro- processor	x		Continous path	Pendant taught CAD/CAM option	Rooting drilling wet-jet adhesives	+/- .010"

continues

| Manufacturer | Model | Payload | | No. | Axes | Powered By |
		Lbs.	Kgs.		Action	
Thermwood	Cartesian 5 Electric Robot TC Series	10-150	45-68	3-4	Cartesian	Elec.
	LS Series	10-25	4.5-11.3	3-4	Cartesian	Elec.
Trallfa (Norway) US Representative: DeVilbiss Toledo, OH (419)470-2021	Trallfa 450	30	13.6	5	Jointed-arm spherical coordinate	Hydraulic
Trallfa	Trallfa 3500	30	13.6	5	Jointed-arm spherical coordinate	Hydraulic
Trallfa	Trallfa 4500	30	13.6	5	Jointed-arm spherical coordinate	Hydraulic
Unimation/Westinghouse Danbury, CT Farmington Hills, MI office (313)478-7780	Unimate* 1000A	50	22.9	3-5	Spherical coordinate	Hydraulic air
Unimation	Unimate* 2000B and 2100	300	136	3-6	Spherical coordinate	Hydraulic
Unimation	Unimate* 2571BB 2671BB	225	136	5-6	Spherical coordinate	Hydraulic
Unimation	Unimate* 4000A 4571BA 4671BA	450	205	5-6	Spherical coordinate	Hydraulic
Unimation	PUMA 260	2	0.9	6	Jointed-arm revolute	Elec.

Control Systems	S E R V O	N O N S E R V O	Sequence of Motions Taught	Programming Method	Major Applications	Repeatable Placement Accuracy
Microprocessor	x		Continuous path	Pendant taught CAD/CAM option	Assembly, deburr load/unload	+/- .020"
Micro-	x		Point-to-point	Pendant	Pick & place	+/- .005"
Minicomputer (30 programs)	x		Continuous path	Manual teach playback	Coating, adhesives, spray painting	+/- .08"
Minicomputer (30 programs)	x		Continuous path	Manual teach playback	Coating, adhesives, spray painting	+/- .08"
Minicomputer (999 programs)	x		Continuous path	Manual teach playback	Coating, adhesives, spray painting	+/- .08"
Special purpose computer	x	x	Point-to-point	Remote pendant	Mchn load/ unload, pick & place, die casting/injection molding	+/- .025"
Special purpose computer	x		Point-to-point	Remote pendant	Mchn load/ unload, spot/ arc welding, mtrl handling	+/- .025" +/- .04"
Minicomputer Val	x		Joint straight line/circular interpolated coordinated axis	Remote pendant or off-line	Mchn load/ unload, spot/ arc welding, mtrl handling	+/- .025"
Special purpose Computer/ Val	x		Point-to-point or joint straight line/ circular interpolated coordinated axis	Remote pendant or off-line	Mchn load/ unload, spot welding	+/- .04"
MiniComputer Val I Val II	x		Joint, straight line/circular interpolated coordinated axis	Remote pendant or off-line	Small parts handling & assembly	+/- .02"

continues

| Manufacturer | Model | Payload | | No. | Axes | Powered By |
		Tbs.	Kgs.		Action	
Unimation	PUMA 550*/560	6.5	2.5	5-6	Jointed-arm revolute	Elect.
Unimation	PUMA 760*	22	10	6	Jointed-arm revolute	Elect.
Unimation	100	10	5	4	Jointed-arm	Elec.
Unimation	Dual Arm	100	45	9	Modified polar	Hydraulic
Unimation (Westinghouse)	6000	100-300 300	45-135	5-3	Rectilinear coordinate	Elec.
Unimation (Westinghouse)	7000	44	20	5	Rectilinear coordinate	Elec.
Yaskawa (Japan) (Represented by: Hobart Brothers)	Motoman* L-10W	22	10	5-6	Jointed-arm spherical coordinate	Elec.
Yaskawa	Motoman L-3	6.6	3	5	Jointed-arm spherical coordinate	Elec.
United States Robots Farmington Hills, MI Office (31)476-1884	Maker* 100	5	2.5	5	Jointed-arm spherical coordinate	Elec.
VSI Automation Troy, MI (313)588-1255	Charley 2*	13.2	6	4	Horizontal Jointed-arm cylindrical coordinate	Elec.
VSI Automation	Charley 4*	22	10	4	Horizontal Jointed-arm cylindrical coordinate	Elec.
VSI Automation	Charley 6*	66	30	4	Horizontal Jointed-arm cylindrical coordinate	Elec.

Control Systems	S E R V O	N O N S E R V O	Sequence of Motions Taught	Pro- gramming Method	Major Applications	Repeatable Placement Accuracy
Mini- computer Val I Val II	x		Joint, straight line/cir- cular interpolated coordinated axis	Remote pendant or off-line	Assembly, parts handling, arc welding, adhesive dispensing, inspection	+/- .004"
Mini- computer Val I Val II	x		Joint, straight line/cir- cular interpolated coordinated axis	remote pendant or off-line	Arc welding, adhesive dispensing, inspection	+/- .008"
PC	x		Point-to-point	Remote pendant	Assembly	+/- .004"
Special purpose computer	x		Point-to-point	Remote pendant	Mtrl handling	+/- .04"
Mini- computer Val II or CNC	x		Coordinated axis	Remote pendant or off-line	Arc welding, mtrl handling and burning	+/- .005"
Micro- computer	x		Continuous path	Remote pendant or off-line	Adaptive control arc welding	+/- .016"
Micro- computer	x		Point-to-point	Remote pendant & off-line	Arc welding, sealer	+/- .008"
Computer	x		Point-to-point	Manual teach playback	Arc welding and light handling	+/- .004"
Micro- processor based	x		Point-to-point continuous path	Remote pendant	Assembly, small parts and mchn load	+/- .002"
Micro- processor	x		Point-to-point	Remote pendant	Assembly	+/- .002"
Micro- processor	x		Point-to-point	Remote pendant	Assembly	+/- .002"
Micro- processor	x		Point-to-point	Remote pendant	Assembly	+/- .002"

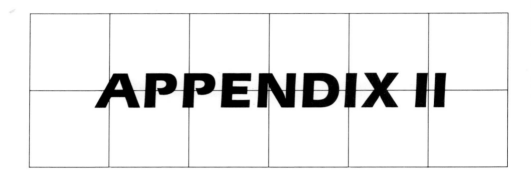

APPENDIX II

☐ **ELECTRONICS AND FLUID POWER SYMBOLS**

Appendix II reproduced from *Robotics*, pp. 247–254, by James W. Masterson, Elmer C. Poe, and Stephen W. Fardo, © 1985 by Reston Publishing Co., Inc., Reston, VA, with permission of Prentice-Hall, Inc., Englewood Cliffs, N.J.

ELECTRONICS SYMBOLS

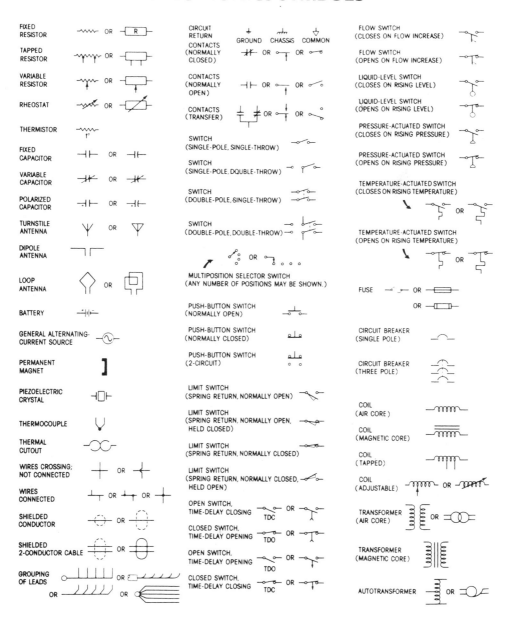

FIXED RESISTOR	—⩗⩗— OR —[R]—
TAPPED RESISTOR	
VARIABLE RESISTOR	
RHEOSTAT	
THERMISTOR	
FIXED CAPACITOR	—⊢— OR —⊣⊢
VARIABLE CAPACITOR	
POLARIZED CAPACITOR	
TURNSTILE ANTENNA	
DIPOLE ANTENNA	
LOOP ANTENNA	
BATTERY	
GENERAL ALTERNATING-CURRENT SOURCE	
PERMANENT MAGNET	
PIEZOELECTRIC CRYSTAL	
THERMOCOUPLE	
THERMAL CUTOUT	
WIRES CROSSING; NOT CONNECTED	
WIRES CONNECTED	
SHIELDED CONDUCTOR	
SHIELDED 2-CONDUCTOR CABLE	
GROUPING OF LEADS	

CIRCUIT RETURN	GROUND CHASSIS COMMON
CONTACTS (NORMALLY CLOSED)	
CONTACTS (NORMALLY OPEN)	
CONTACTS (TRANSFER)	
SWITCH (SINGLE-POLE, SINGLE-THROW)	
SWITCH (SINGLE-POLE, DOUBLE-THROW)	
SWITCH (DOUBLE-POLE, SINGLE-THROW)	
SWITCH (DOUBLE-POLE, DOUBLE-THROW)	
MULTIPOSITION SELECTOR SWITCH (ANY NUMBER OF POSITIONS MAY BE SHOWN.)	
PUSH-BUTTON SWITCH (NORMALLY OPEN)	
PUSH-BUTTON SWITCH (NORMALLY CLOSED)	
PUSH-BUTTON SWITCH (2-CIRCUIT)	
LIMIT SWITCH (SPRING RETURN, NORMALLY OPEN)	
LIMIT SWITCH (SPRING RETURN, NORMALLY OPEN, HELD CLOSED)	
LIMIT SWITCH (SPRING RETURN, NORMALLY CLOSED)	
LIMIT SWITCH (SPRING RETURN, NORMALLY CLOSED, HELD OPEN)	
OPEN SWITCH, TIME-DELAY CLOSING	TDC
CLOSED SWITCH, TIME-DELAY OPENING	TDO
OPEN SWITCH, TIME-DELAY OPENING	TDO
CLOSED SWITCH, TIME-DELAY CLOSING	TDC

FLOW SWITCH (CLOSES ON FLOW INCREASE)	
FLOW SWITCH (OPENS ON FLOW INCREASE)	
LIQUID-LEVEL SWITCH (CLOSES ON RISING LEVEL)	
LIQUID-LEVEL SWITCH (OPENS ON RISING LEVEL)	
PRESSURE-ACTUATED SWITCH (CLOSES ON RISING PRESSURE)	
PRESSURE-ACTUATED SWITCH (OPENS ON RISING PRESSURE)	
TEMPERATURE-ACTUATED SWITCH (CLOSES ON RISING TEMPERATURE)	OR
TEMPERATURE-ACTUATED SWITCH (OPENS ON RISING TEMPERATURE)	OR
FUSE	—⊸~⊸ OR —▭— OR —▱—
CIRCUIT BREAKER (SINGLE POLE)	
CIRCUIT BREAKER (THREE POLE)	
COIL (AIR CORE)	
COIL (MAGNETIC CORE)	
COIL (TAPPED)	
COIL (ADJUSTABLE)	OR
TRANSFORMER (AIR CORE)	OR
TRANSFORMER (MAGNETIC CORE)	
AUTOTRANSFORMER	OR

continues

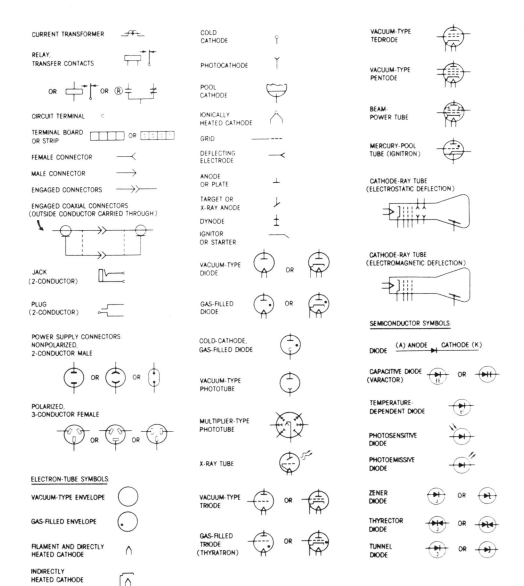

CURRENT TRANSFORMER

RELAY,
TRANSFER CONTACTS

OR   OR ®

CIRCUIT TERMINAL

TERMINAL BOARD
OR STRIP   OR

FEMALE CONNECTOR

MALE CONNECTOR

ENGAGED CONNECTORS

ENGAGED COAXIAL CONNECTORS
(OUTSIDE CONDUCTOR CARRIED THROUGH)

JACK
(2-CONDUCTOR)

PLUG
(2-CONDUCTOR)

POWER SUPPLY CONNECTORS
NONPOLARIZED,
2-CONDUCTOR MALE

OR   OR

POLARIZED,
3-CONDUCTOR FEMALE

OR   OR

ELECTRON-TUBE SYMBOLS

VACUUM-TYPE ENVELOPE

GAS-FILLED ENVELOPE

FILAMENT AND DIRECTLY
HEATED CATHODE

INDIRECTLY
HEATED CATHODE

COLD
CATHODE

PHOTOCATHODE

POOL
CATHODE

IONICALLY
HEATED CATHODE

GRID

DEFLECTING
ELECTRODE

ANODE
OR PLATE

TARGET OR
X-RAY ANODE

DYNODE

IGNITOR
OR STARTER

VACUUM-TYPE
DIODE   OR

GAS-FILLED
DIODE   OR

COLD-CATHODE,
GAS-FILLED DIODE

VACUUM-TYPE
PHOTOTUBE

MULTIPLIER-TYPE
PHOTOTUBE

X-RAY TUBE

VACUUM-TYPE
TRIODE   OR

GAS-FILLED
TRIODE   OR
(THYRATRON)

VACUUM-TYPE
TEDRODE

VACUUM-TYPE
PENTODE

BEAM-
POWER TUBE

MERCURY-POOL
TUBE (IGNITRON)

CATHODE-RAY TUBE
(ELECTROSTATIC DEFLECTION)

CATHODE-RAY TUBE
(ELECTROMAGNETIC DEFLECTION)

SEMICONDUCTOR SYMBOLS

DIODE   (A) ANODE   CATHODE (K)

CAPACITIVE DIODE   OR
(VARACTOR)

TEMPERATURE-
DEPENDENT DIODE

PHOTOSENSITIVE
DIODE

PHOTOEMISSIVE
DIODE

ZENER
DIODE   OR

THYRECTOR
DIODE   OR

TUNNEL
DIODE   OR

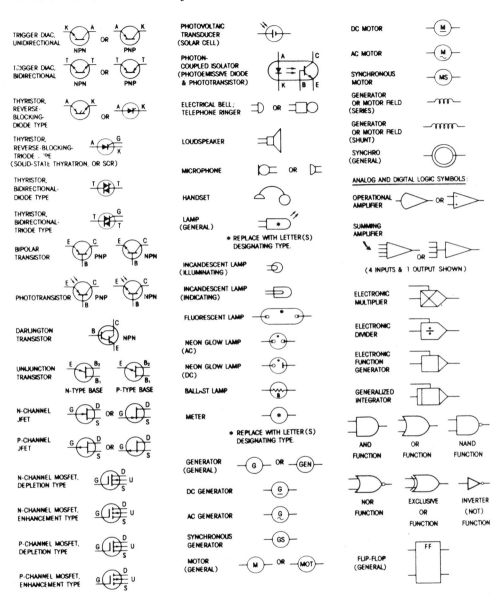

TRIGGER DIAC, UNIDIRECTIONAL

TRIGGER DIAC, BIDIRECTIONAL

THYRISTOR, REVERSE-BLOCKING-DIODE TYPE

THYRISTOR, REVERSE-BLOCKING-TRIODE TYPE (SOLID-STATE THYRATRON, OR SCR)

THYRISTOR, BIDIRECTIONAL-DIODE TYPE

THYRISTOR, BIDIRECTIONAL-TRIODE TYPE

BIPOLAR TRANSISTOR

PHOTOTRANSISTOR

DARLINGTON TRANSISTOR

UNIJUNCTION TRANSISTOR

N-CHANNEL JFET

P-CHANNEL JFET

N-CHANNEL MOSFET, DEPLETION TYPE

N-CHANNEL MOSFET, ENHANCEMENT TYPE

P-CHANNEL MOSFET, DEPLETION TYPE

P-CHANNEL MOSFET, ENHANCEMENT TYPE

PHOTOVOLTAIC TRANSDUCER (SOLAR CELL)

PHOTON-COUPLED ISOLATOR (PHOTOEMISSIVE DIODE & PHOTOTRANSISTOR)

ELECTRICAL BELL; TELEPHONE RINGER

LOUDSPEAKER

MICROPHONE

HANDSET

LAMP (GENERAL)

* REPLACE WITH LETTER(S) DESIGNATING TYPE.

INCANDESCENT LAMP (ILLUMINATING)

INCANDESCENT LAMP (INDICATING)

FLUORESCENT LAMP

NEON GLOW LAMP (AC)

NEON GLOW LAMP (DC)

BALLAST LAMP

METER

* REPLACE WITH LETTER(S) DESIGNATING TYPE.

GENERATOR (GENERAL)

DC GENERATOR

AC GENERATOR

SYNCHRONOUS GENERATOR

MOTOR (GENERAL)

DC MOTOR

AC MOTOR

SYNCHRONOUS MOTOR

GENERATOR OR MOTOR FIELD (SERIES)

GENERATOR OR MOTOR FIELD (SHUNT)

SYNCHRO (GENERAL)

ANALOG AND DIGITAL LOGIC SYMBOLS:

OPERATIONAL AMPLIFIER

SUMMING AMPLIFIER

(4 INPUTS & 1 OUTPUT SHOWN)

ELECTRONIC MULTIPLIER

ELECTRONIC DIVIDER

ELECTRONIC FUNCTION GENERATOR

GENERALIZED INTEGRATOR

AND FUNCTION

OR FUNCTION

NAND FUNCTION

NOR FUNCTION

EXCLUSIVE OR FUNCTION

INVERTER (NOT) FUNCTION

FLIP-FLOP (GENERAL)

FLUID POWER SYMBOLS

FLUID CONDUCTORS

WORKING LINE
(MAIN)

PILOT LINE
(FOR CONTROL)

EXHAUST AND
LIQUID DRAIN LINE

FLOW DIRECTION,
HYDRAULIC

FLOW DIRECTION,
PNEUMATIC

LINE WITH
FIXED RESTRICTION

FLEXIBLE
LINE

QUICK DISCONNECT
WITHOUT CHECKS

CONNECTED   DISCONNECTED

QUICK DISCONNECT
WITH ONE CHECK

CONNECTED   DISCONNECTED

QUICK DISCONNECT
WITH TWO CHECKS

CONNECTED   DISCONNECTED

ENERGY AND FLUID STORAGE

VENTED
RESERVOIR

PRESSURIZED
RESERVOIR

RESERVOIR WITH
CONNECTING LINES

ABOVE FLUID LEVEL

BELOW FLUID LEVEL

SPRING-LOADED
ACCUMULATOR

GAS-CHARGED
ACCUMULATOR

WEIGHTED
ACCUMULATOR

RECEIVER FOR AIR
OR OTHER GASES

FLUID CONDITIONERS

HEATER

INSIDE TRIANGLES INDICATE
THE INTRODUCTION OF HEAT

HEATER

OUTSIDE TRIANGLES INDICATE
A LIQUID HEATING MEDIUM

HEATER

OUTSIDE TRIANGLES INDICATE
A GASEOUS HEATING MEDIUM

COOLER   OR

INSIDE TRIANGLES INDICATE
HEAT DISSIPATION

COOLER   OR

OUTSIDE TRIANGLES INDICATE
A LIQUID OR GASEOUS COOLING MEDIUM

TEMPERATURE
CONTROLLER   OR

OUTSIDE TRIANGLES INDICATE A
LIQUID OR GASEOUS MEDIUM

FILTER
OR STRAINER

SEPARATOR
WITH MANUAL DRAIN

SEPARATOR
WITH AUTOMATIC DRAIN

FILTER-SEPARATOR
WITH MANUAL DRAIN

FILTER-SEPARATOR
WITH AUTOMATIC DRAIN

DESSICATOR
(CHEMICAL DRYER)

LUBRICATOR
WITHOUT DRAIN

LUBRICATOR
WITH MANUAL DRAIN

LINEAR DEVICES

SINGLE-ACTING CYLINDERS
(HYDRAULIC AND PNEUMATIC)

DOUBLE-ACTING CYLINDER
WITH SINGLE END ROD

DOUBLE-ACTING CYLINDER
WITH DOUBLE END ROD

PRESSURE
INTENSIFIER

HYDRAULIC   PNEUMATIC

SERVO POSITIONER

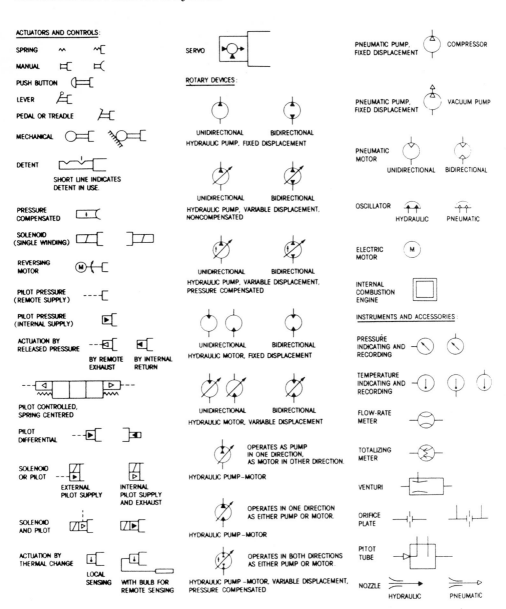

ACTUATORS AND CONTROLS:

SPRING

MANUAL

PUSH BUTTON

LEVER

PEDAL OR TREADLE

MECHANICAL

DETENT

SHORT LINE INDICATES
DETENT IN USE.

PRESSURE
COMPENSATED

SOLENOID
(SINGLE WINDING)

REVERSING
MOTOR

PILOT PRESSURE
(REMOTE SUPPLY)

PILOT PRESSURE
(INTERNAL SUPPLY)

ACTUATION BY
RELEASED PRESSURE

BY REMOTE   BY INTERNAL
EXHAUST   RETURN

PILOT CONTROLLED,
SPRING CENTERED

PILOT
DIFFERENTIAL

SOLENOID
OR PILOT

EXTERNAL   INTERNAL
PILOT SUPPLY   PILOT SUPPLY
                AND EXHAUST

SOLENOID
AND PILOT

ACTUATION BY
THERMAL CHANGE

LOCAL
SENSING   WITH BULB FOR
               REMOTE SENSING

SERVO

ROTARY DEVICES:

UNIDIRECTIONAL   BIDIRECTIONAL
HYDRAULIC PUMP, FIXED DISPLACEMENT

UNIDIRECTIONAL   BIDIRECTIONAL
HYDRAULIC PUMP, VARIABLE DISPLACEMENT,
NONCOMPENSATED

UNIDIRECTIONAL   BIDIRECTIONAL
HYDRAULIC PUMP, VARIABLE DISPLACEMENT,
PRESSURE COMPENSATED

UNIDIRECTIONAL   BIDIRECTIONAL
HYDRAULIC MOTOR, FIXED DISPLACEMENT

UNIDIRECTIONAL   BIDIRECTIONAL
HYDRAULIC MOTOR, VARIABLE DISPLACEMENT

OPERATES AS PUMP
IN ONE DIRECTION,
AS MOTOR IN OTHER DIRECTION.

HYDRAULIC PUMP-MOTOR

OPERATES IN ONE DIRECTION
AS EITHER PUMP OR MOTOR.

HYDRAULIC PUMP-MOTOR

OPERATES IN BOTH DIRECTIONS
AS EITHER PUMP OR MOTOR.

HYDRAULIC PUMP-MOTOR, VARIABLE DISPLACEMENT,
PRESSURE COMPENSATED

PNEUMATIC PUMP,   COMPRESSOR
FIXED DISPLACEMENT

PNEUMATIC PUMP,   VACUUM PUMP
FIXED DISPLACEMENT

PNEUMATIC
MOTOR
          UNIDIRECTIONAL   BIDIRECTIONAL

OSCILLATOR
          HYDRAULIC   PNEUMATIC

ELECTRIC   M
MOTOR

INTERNAL
COMBUSTION
ENGINE

INSTRUMENTS AND ACCESSORIES:

PRESSURE
INDICATING AND
RECORDING

TEMPERATURE
INDICATING AND
RECORDING

FLOW-RATE
METER

TOTALIZING
METER

VENTURI

ORIFICE
PLATE

PITOT
TUBE

NOZZLE
          HYDRAULIC   PNEUMATIC

continues

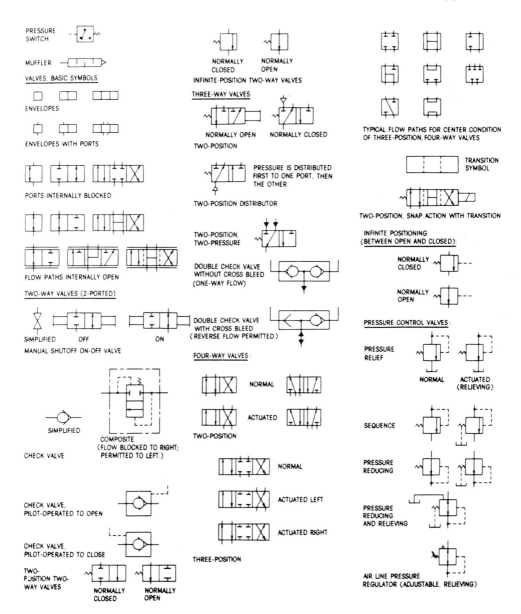

PRESSURE
SWITCH

MUFFLER

VALVES, BASIC SYMBOLS

ENVELOPES

ENVELOPES WITH PORTS

PORTS INTERNALLY BLOCKED

FLOW PATHS INTERNALLY OPEN

TWO-WAY VALVES (2-PORTED)

SIMPLIFIED   OFF   ON

MANUAL SHUTOFF ON-OFF VALVE

SIMPLIFIED

CHECK VALVE

COMPOSITE
(FLOW BLOCKED TO RIGHT;
PERMITTED TO LEFT.)

CHECK VALVE,
PILOT-OPERATED TO OPEN

CHECK VALVE,
PILOT-OPERATED TO CLOSE

TWO-
POSITION TWO-
WAY VALVES

NORMALLY   NORMALLY
CLOSED   OPEN

NORMALLY   NORMALLY
CLOSED   OPEN

INFINITE-POSITION TWO-WAY VALVES

THREE-WAY VALVES:

NORMALLY OPEN   NORMALLY CLOSED

TWO-POSITION

PRESSURE IS DISTRIBUTED
FIRST TO ONE PORT, THEN
THE OTHER

TWO-POSITION DISTRIBUTOR

TWO-POSITION,
TWO-PRESSURE

DOUBLE CHECK VALVE
WITHOUT CROSS BLEED
(ONE-WAY FLOW)

DOUBLE CHECK VALVE
WITH CROSS BLEED
(REVERSE FLOW PERMITTED)

FOUR-WAY VALVES:

NORMAL

ACTUATED

TWO-POSITION

NORMAL

ACTUATED LEFT

ACTUATED RIGHT

THREE-POSITION

TYPICAL FLOW PATHS FOR CENTER CONDITION
OF THREE-POSITION, FOUR-WAY VALVES

TRANSITION
SYMBOL

TWO-POSITION, SNAP ACTION WITH TRANSITION

INFINITE POSITIONING
(BETWEEN OPEN AND CLOSED):

NORMALLY
CLOSED

NORMALLY
OPEN

PRESSURE CONTROL VALVES:

PRESSURE
RELIEF

NORMAL   ACTUATED
            (RELIEVING)

SEQUENCE

PRESSURE
REDUCING

PRESSURE
REDUCING
AND RELIEVING

AIR LINE PRESSURE
REGULATOR (ADJUSTABLE, RELIEVING)

INFINITE POSITIONING VALVES:

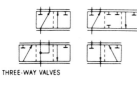

THREE-WAY VALVES

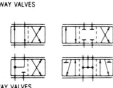

FOUR-WAY VALVES

FLOW-CONTROL VALVES:

ADJUSTABLE, NONCOMPENSATED
(FLOW CONTROL IN EACH DIRECTION)

ADJUSTABLE,
WITH BYPASS

FLOW CONTROLLED TO RIGHT,
FLOW TO LEFT BYPASSES CONTROL.

ADJUSTABLE AND PRESSURE
COMPENSATED, WITH BYPASS

ADJUSTABLE, TEMPERATURE
AND PRESSURE COMPENSATED

AIR LINE ACCESSORIES:

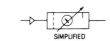

SIMPLIFIED

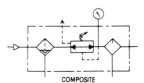

COMPOSITE

FILTER, REGULATOR, AND LUBRICATOR

GLOSSARY

accuracy degree to which a robot can place an object in a given spot repeatedly

actuator motors, cylinders, or other mechanisms used to power robots

articulation ability of a robot to extend and retract, swing or rotate, and elevate its arm

ASCII code used to transfer information from a keyboard to the processor, MPU, or CPU

assemble to put together

backlash looseness in the gears where they mesh

ball screw method of using ball bearings to substitute for screw threads (The ball screw changes rotary motion to linear motion.)

brushless motor DC motor that operates without brushes (An electronic circuit controls its field excitation.)

Cartesian coordinate simplest of the coordinates because it refers to up-and-down, back-and-forth, and in-and-out movement of a robot arm

chip an integrated circuit that contains many transistors, diodes, and resistors to make up various electronic circuits on a small (about 8 mm square) piece of silicon; device that stores the computer's program and acts as a memory

CIM   computer-integrated manufacturing; one of a number of proposed organizational methods for manufacturing products in large quantities

collision avoidance   ability of a robot to avoid colliding with the part it is supposed to pick up

contact sensor   sensor that detects the presence of an object by actually making contact

controller   unit needed to control the robot

coordinates   points and planes in reference to the movement capability of a robot arm

CPU   central processing unit

dead zone   safety zone where robot arm does not move during normal operation

degrees of freedom   ability of a robot to move in six axes

die casting   use of dies to form hot metal into desired shapes

end effector   device mounted on end of manipulator or robot arm

fabricate   to make something

feedback   ability of a device to feed a signal back to its controller to aid in keeping track of the position of the manipulator or gripper

flow line transfer   robots designed to pick up two or more pieces at a time and transfer them off a machining line onto a second transfer line located parallel to the first one

gripper   located on end of manipulator arm and used to pick up things

hard automation   use of conventional assembly line method of producing a manufactured product with dedicated equipment

harmonic drive   type of drive that uses a flexspline, circular spline, and wave generator to accurately position a manipulator with no backlash and little noise

hydraulics   use of pressure on a fluid to drive an end effector or a manipulator

interface   proper connections between a robot and its programmed computer or microprocessor

interfacing   matching up of a device with a computer or microprocessor so they operate as one unit

ladder diagram   drawing of the electrical circuit of the sequence of switches needed to cause a robot to perform programmed activities

lane loader   pick-and-place robot used to adjust the feed between fast and slow or slow and fast lines

language method of speaking to a robot (Languages used are VAL, AL, AML, Pascal, and ADA.)

lead through leading the continuous-path robot manipulator through a sequence of motions to accomplish a task

LED light-emitting diode

LERT classification system for robots based on four basic motion capabilities

limit switch switch designed to be used with a moving body that should not go past a given point

line tracking having the robot keep up and move along with the production line, doing its work as it moves with the line

machine vision system system that allows the robot to recognize and verify parts

manipulator one of three basic parts of a robot

microcomputer small-scale computer

MIG type of metal-in-gas welding where an inert gas surrounds the welding spot or area while it is in the molten state

minicomputer mid-size computer, slightly larger than a microcomputer and smaller than a mainframe

microprocessor electronic device consisting of IC chips that have memory and the ability to generate signals to control a robot

MPU microprocessor unit

MVS machine vision system; method of giving the robot the ability to see

object recognition ability of a robot to recognize certain shapes and choose the right one for processing

off-line refers to programming of a robot by use of a computer and then placing the program in the robot's controller for action

palletizing task in which a robot stacks parts or boxes on a pallet

parallel port method for connecting computers and peripheral devices so they can share data (uses eight or more wires)

plugging method of stopping an electric motor by reversing the polarity of its power source

pneumatic drive using air pressure for driving the manipulator

power supply supplies power to the robot

positioning ability of a robot to place an object in a desired location

process flow orderly flow of parts and materials to keep production going

program   sequence of commands instructing a robot to perform some task

programmable robot   robot that can be programmed or taught with a teach box, a keyboard, or another input device

programmer   person who teaches the robot; person who can communicate with the robot in its language

proximity sensor   device used to detect how close an object is

pumps   devices, usually electrically driven, used to increase the pressure on a hydraulic fluid

range sensor   device used to detect the precise distance between an object and the gripper or manipulator

relay   electromechanical device that is energized and closes or opens switches (It has a coil that can be magnetized by the presence of an electric current.)

repeatability   ability of a robot to place an object in the same spot repeatedly

robot   a system that simulates human activities from computer instruction

roller chain   same type of chain used in bicycles; used to drive manipulators and end effectors

RS-232C standard   standard that uses -3 to -25 volts for logic 1 and $+3$ to $+25$ volts for logic 0

sensor   device used to detect changes in temperature, light, pressure, sound, and other functions needed to make a robot aware of various conditions

serial port   method for connecting computers and peripheral devices so they can share data over distances of fifty or more feet (uses two wires)

servo motors   motors driven by signals rather than by straight power line voltage and current; motors whose driving signal is a function of the difference between command position and/or rate and measured actual position and/or rate

60 mA standard   uses 60 mA for logic 1 and zero for logic 0

software   information programmed on floppy disks, hard disks, magnetic tape, or drums

solenoid   coil of wire with a plunger that can turn on or off a fluid or air line

spot welding   welding a small spot between two electrodes of a spot welder

stepper motor   DC motor whose rotor can be made to turn as little as 1.8°

strain gage   device made of thin wire that reacts to stretching of the wire; made of semiconductor materials today

synchronous belt   type of belt that has teeth that fit a pulley with grooves so it does not slip

tactile sensor   device used to detect the presence of an object by touch

target   point at which the arm is expected to reach for picking up an object

teach pendant   device used to teach the robot memory a new program

thermistor   device that changes its resistance in the reverse manner from normal (If temperature increases, it lowers its resistance.)

thermocouple   union made by two dissimilar metals (When heated, the junction produces a small electrical current.)

TIG   type of welding (tungsten-in-gas)

transducer   device used to convert mechanical energy to electrical energy

TTL standard   transistor-transistor logic standard that uses a 5 volt signal for logic 1 and 0 for logic 0

20 mA standard   uses 20 mA for logic 1 and zero for logic 0

V-belt   belt shaped to fit a V pulley and used to drive a manipulator

work envelope   space in which the robot arm moves during its normal work cycle

worm gears   change linear motion to rotary motion or vice versa

BIBLIOGRAPHY

Albus, J. S. *Brains, Behavior, and Robotics*. Peterborough, NH: BYTE Books, 1981.

Baar, A., and E. A. Feigenbaum. *The Handbook of Artificial Intelligence*. Stanford University, Stanford, CA: Heurish Tech Press, 1981.

D'Ignazio, Fred. *Working Robots*. Hasbrouck Heights, NJ: Hayden Book Company, 1984.

Duffy, Joseph. *Analysis of the Mechanisms of Robot Manipulators*. New York: John Wiley & Sons, Inc., 1980.

Engelberger, J. *Robotics in Practice, Management and Applications of Industrial Robots*. New York: AMACOM, 1981.

Glorioso, R. M., and F. C. Colon Osorio. *Engineering Intelligent Systems*. Bedford, MA: Digital Press, 1980.

Heath, Larry. *Fundamentals of Robotics*. Reston, VA: Reston Publishing Co., 1985.

Hoekstra, Robert L. *Robotics and Automated Systems*. Cincinnati, OH: South-Western Publishing Co., 1986.

Holland, John M. *Basic Robotics Concepts*. Indianapolis, IN: Howard W. Sams, Inc., 1983.

Hunt, V. Daniel. *Industrial Robotics Handbook*. New York: Industrial Press, Inc., 1983.

Iver, David F., and Robert W. Bolz. *Robotics Sourcebook and Dictionary*. New York: Industrial Press, Inc., 1983.

Kafrissen, Edward K., and Mark Stephans. *Industrial Robots and Robotics.* Reston, VA: Reston Publishing Company, 1984.

Malcolm, Douglas R., Jr. *Robotics: An Introduction.* Boston, MA: Breton Publishers, 1985.

Masterson, James W., Elmer C. Poe, and Stephen W. Fardo. *Robotics.* Reston, VA: Reston Publishing Company, 1985.

Orin, D. E. "Supervisory Control of a Multilegged Robot," *International Journal of Robotics Research*, vol. 1, no. 1, Spring 1982.

Paul, Richard P. *Robotic Manipulators.* Cambridge, MA: MIT Press, 1981.

Snyder, Wesley E. *Industrial Robots: Computer Interfacing and Control.* Englewood Cliffs, NJ: Prentice-Hall, 1985.

Weinstein, Martin B. *Android Design.* Rochelle Park, NJ: Hayden Book Co., 1981.

Winston, P. H. *Artificial Intelligence.* Reading, MA: Addison-Wesley Publishing Company, 1981.

INDEX